Robotics and Smart Autonomous Systems

The text discusses fundamental, advanced concepts and applications of robotics and autonomous systems. It further discusses important topics, such as robotic techniques in the manufacturing sector, applications of smart autonomous systems in the healthcare sector, resource optimization in mobile robotics, and smart autonomous transport systems.

Features

- Covers design and application aspects of robotic systems for implementing the concepts of smart manufacturing with reduced human intervention, better accuracy, and enhanced production capacity.
- Discusses techniques including supervised learning, unsupervised learning, and reinforced learning with real-life examples.
- Highlights a unified intermodal approach for automated transportation, including cars, trucks, ships, and port management.
- Explains the mechanical design of planetary rovers, and the mechanical design of space manipulators, actuators, and sensors.
- Presents programming tools and platforms for autonomous robotic systems.

The book is primarily written for senior undergraduates, graduate students, and academic researchers in fields including electrical engineering, electronics and communications engineering, computer science and engineering, and automotive engineering.

Wireless Communications and Networking Technologies: Classifications, Advancement and Applications
Series Editors: D.K. Lobiyal, R.S. Rao, and Vishal Jain

The series addresses different algorithms, architecture, standards and protocols, tools and methodologies that could be beneficial in implementing a next-generation mobile network for communication. Aimed at senior undergraduate students, graduate students, academic researchers, and professionals, the proposed series focuses on the fundamentals and advances of wireless communication and networking, such as mobile ad-hoc networks (MANET), wireless sensor networks (WSN), wireless mess networks (WMN), vehicular ad-hoc networks (VANET), vehicular cloud networks (VCN), vehicular sensor networks (VSN), reliable cooperative networks (RCN), mobile opportunistic networks (MON), delay tolerant networks (DTN), flying ad-hoc networks (FANET), and wireless body sensor networks (WBSN).

Wireless Communication: Advancements and Challenges
Prashant Ranjan, Ram Shringar Rao, Krishna Kumar and Pankaj Sharma

Wireless Communication with Artificial Intelligence: Emerging Trends and Applications
Anuj Singal, Sandeep Kumar, Sajjan Singh and Ashish Kr. Luhach

Computational Intelligent Security in Wireless Communications
Suhel Ahmad Khan, Rajeev Kumar, Omprakash Kaiwartya, Raees Ahmad Khan and Mohammad Faisal

Networking Technologies in Smart Healthcare: Innovations and Analytical Approaches
Pooja Singh, Omprakash Kaiwartya, Nidhi Sindhwani, Vishal Jain and Rohit Anand

Artificial Intelligence in Cyber Physical Systems: Principles and Applications
Anil Kumar Sagar, Parma Nand, Neetesh Kumar, Sanjoy Das and Subrata Sahana

For more information about this series, please visit: https://www.routledge.com/Wireless%20Communications%20and%20Networking%20Technologies/book-series/WCANT

Robotics and Smart Autonomous Systems

Technology and Applications

Edited by

Rashmi Priyadarshini, Ram Mohan Mehra, Amit Sehgal, and Prabhu Jyot Singh

CRC Press
Taylor & Francis Group
Boca Raton London New York

CRC Press is an imprint of the
Taylor & Francis Group, an **informa** business

Designed cover image: © Shutterstock

First edition published 2025
by CRC Press
2385 NW Executive Center Drive, Suite 320, Boca Raton FL 33431

and by CRC Press
4 Park Square, Milton Park, Abingdon, Oxon, OX14 4RN

CRC Press is an imprint of Taylor & Francis Group, LLC

ISBN: 978-1-032-62770-0 (hbk)
ISBN: 978-1-032-87409-8 (pbk)
ISBN: 978-1-003-53708-3 (ebk)

DOI: 10.1201/9781003537083

Typeset in Sabon
by Newgen Publishing UK

Contents

Preface

The book entitled "Robotics and Smart Autonomous Systems: Technology and Applications" consists of chapters that provide a valuable resource for researches, scientists, professionals, academicians, and students dealing with the new challenges and advances in the area of robotics and smart autonomous systems. It also provides an approach from the design perspective of such systems in various fields of application. Robotics has played a major role in transforming traditional manufacturing into smart and computer-integrated manufacturing leading to Industry 4.0. In addition to stationary robotic systems used in industries, various mobile robots are being developed. Autonomous vehicles, humanoids, ornithopters, and other biologically inspired robots are already present and are becoming a part of our society. Intelligence in robotic systems is a highly desired feature, empowering robots to sense, process, and make choices to take actions. The book also focuses on various techniques and processes embedding intelligence in robotic systems.

Robotics and Smart Autonomous Systems: Technology and Applications is a textbook about the science and technology which provides a plan to build and maintain an AI-enabled robotics system. It includes easy-to-follow concepts of robotics and AI systems. We wrote this book because we are excited about the emerging applications and innovations of AI as an integrated science. AI applications and innovations for robots have a coherent, formal theory and a rambunctious experimental wing.

Rich in illustrations with over 80 photos and nearly 250 pages, the book gives an insight into the successful construction of AI-controlled robots. The book is written in simple language for the easy comprehension of students and faculty members.

The book can be used as a source to understand the basic concepts of robotics, designing of AI-integrated robots, the role of AI in smart robotic systems, programming platforms of autonomous robotic systems, applications of smart autonomous systems in healthcare, robots in agriculture – farm mechanization and precision farming, innovations in STEM education – robotics and AI, and robots in space exploration.

The applications of AI in different application domains will help readers to understand the actual use of AI in real-life examples.

Predictions for the state of AI and robotics in 2025 would assert that AI and robotics will be integrated into nearly every aspect of most people's daily lives.

Besides, the book also contains methods that allow the cost of new robotic systems controlled by AI to be calculated. Pictorial demonstrations have been included for easy comprehension.

The book is easy to follow and understand and is a textbook at an undergraduate and postgraduate level.

It includes problems with solutions, exercises for practice, where some derivations that are too lengthy for the book are also included.

While the focus is on advanced and recent applications, efforts have been made to explain the fundamental concepts in an easily understandable way with relevant references and examples.

The references listed at the end of each chapter are not meant to be a comprehensive inventory, rather these references, both classic works and more recent surveys, have been directly used in this book. Thus the list provides an excellent overview of the related or prerequisite literature.

We hope that no student, faculty, or researcher feels slighted by any omission and we will be very happy to receive feedback if someone feels that an idea has not been appropriately covered.

About the editors

Rashmi Priyadarshini has a Ph.D, M.E, (Electronics and Communication Engineering, from IGIT, GGSIP University, Delhi, India) and B.Tech degree, (Electronics and Communication Engineering, from Institution of Electronics and Telecommunication Engineering, New Delhi, India). She has published 100 research papers in various reputed journals. She has 18 years of teaching experience in various engineering colleges across India. She was associated with Sharda University, India, in 2006, and is currently associated with the School of Engineering & Technology as a Professor in the Department of Electrical Electronics & Communication Engineering. Her main research interests are the design and deployment of wireless sensor networks for assistive purposes and monitoring. She has guided multiple research scholars in the area of wireless sensor networks and memristors. In total, she has eight patents, out of which one is funded by the Govt. of India (MSME). She is a senior Member of IEEE and Faculty IN charge of students' chapter of IETE. She was awarded the best researcher award for her outstanding contributions in the field of wireless sensor networks at the International Conference on Computational Intelligence for Engineering and Management Applications (CIEMA) – 2022. She received the accolade of being the best paper presenter at the 3rd International Conference on Advances in Science, Engineering & Management (ICASEM-2022). She has served as a reviewer for *IEEE Transaction on Wireless Communication* and *Journal of Network and Computer Applications*.

Ram Mohan Mehra holds a Ph.D. degree in Physics from the University of Delhi, Delhi. He superannuated from the University of Delhi in January 2010 and joined as Professor Emeritus at Sharda University. At the University of Delhi, he was engaged in teaching, research, and administration. He has been a post-doctoral research fellow of the Japan Society for Promotion of Sciences (JSPS, Japan). He is formally HOD, Department of ECE, Dean Research. Presently he is an Advisor at the Learning Resource Center, Sharda University. He has already supervised 45 Ph.D. research scholars for the award of doctoral degrees. He is presently supervising six research scholars for Ph.D. degree at Sharda University. He has written two review articles, mainly on amorphous semiconductors, which have been widely accepted and recognized by the international research community. He has published 211 research papers in journals cited in SCOPUS with a h-index of 37 and citations of more than 4000. He is reviewer for research papers of several important international journals, such as *Applied Physics Letters* (AIP, USA), *Journal of Applied Physics* (AIP, USA), *Journal of Non-crystalline Solids* (Elsevier Science), *Thin Solid Films* (Elsevier Science), *Journal of Applied Polymer Science* (Wiley Inter Science, USA), *Solar Energy Materials and Solar Cells* (SOLMAT), *Applied Surface Science* (Elsevier Science), and *Polymer Engineering Science*. He has executed more than 17 research projects sponsored by national (DRDO, UGC, DST, CSIR, MNRE, IUAC) and international (NSF, USA & JSPS, Japan) agencies. He has widely traveled abroad for research, teaching, and consultancy assignments to universities in the USA, Japan, and Europe. He is member of several academic and professional bodies and has chaired the Semiconductor Society (India). He has organized national and international conferences in the field of semiconductors. He has been deputed by the Government of India for development of faculty.

Amit Sehgal has a Ph.D. in Electronics Engineering from Dr. A P J Abdul Kalam Technical University, Lucknow. He completed his M. Tech. in electronics and communication engineering from Guru Nanak Dev Engineering College, Ludhiana and B. E. in electronics and communication engineering from C R State College of Engineering (now DCRUST). He has more than 20 years of experience in industry, academia, and research. He has contributed to several research and development projects in the field of wireless networks, health informatics, and IoT, including with corporate and international academic collaboration. Several projects guided by him have received project grants under various State and Central Government Schemes. He is a voluntary startup mentor for Atal Innovation Mission Incubators and Atal Tinkering Labs. He has been contributing to the startup ecosystem by mentoring innovators and early age startups. On completion of his Cambridge Certification for International Teachers and Trainers and Dale Carnegie Certificate for High Impact Teaching Skills, he made a significant contribution to pedagogic uplift of engineering education by serving as Master Trainer for Mission 10X – a faculty development initiative by Wipro Technologies, for more than 2 years. He has more than 35 research papers in journals and conferences. His research interests include wireless networks, remote monitoring systems, and IoT. He has mentored several student entrepreneurs and has wide experience in the field of innovation and incubation.

Prabhu Jyot Singh serves as a Lecturer at CQ University, Australia, having earned his Ph.D. from the same institution. With a fervent dedication to research and education, he imparts knowledge to a diverse student body spanning undergraduate to postgraduate levels, embracing various social and cultural backgrounds. A passionate and driven academic, Dr. Singh specializes in computer networks, particularly focusing on private UAV networks, communication protocols, LAN of drones and UAV communication, software development, software engineering, artificial intelligence, and robotics. His extensive experience of over 8 years in higher education equips him with a profound understanding of the field, enabling him to effectively teach and mentor students across a spectrum of expertise. His commitment extends to staying abreast of cutting-edge technological advancements, notably in computer networking, UAV networks, software engineering, artificial intelligence, and machine learning, which he seamlessly integrates into his teaching and research endeavors. Dr. Singh's scholarly contributions manifest through numerous publications in esteemed indexed journals, conferences, and book chapters, showcasing his adeptness in conducting high-quality research. His doctoral thesis, "Design and implementation of an experimental UAV network," underscores his keen interest in exploring UAV applications in networking, further solidifying his expertise in this domain.

Contributors

Yamiko Daniel Banda
Sharda University
Greater Noida, Uttar Pradesh, India

Sushant Bhatt
Shri Ramswaroop Memorial University
Lucknow, Uttar Pradesh, India

Pushpa Choudhary
Galgotias University
Greater Noida, Uttar Pradesh, India

T. Jayachitra
Sharda University
Greater Noida, Uttar Pradesh, India

Ngum Breayon Kwe
Sharda University
Greater Noida, Uttar Pradesh, India

Sai K Lakkaraju
Wentworth Institute Of Higher Education
Sydney, New South Wales, Australia

Neeraj Makhija
Scanaptics
Sydney, New South Wales, Australia

Premkumar Mathusudhanan
Scanaptics
Sydney, New South Wales, Australia

Arpana Mishra
Dronacharya Group Of Institutions
Greater Noida, Uttar Pradesh, India

Muhammad Abdullahi Musa
Sharda University
Greater Noida, Uttar Pradesh, India

Rashmi Priyadarshini
Sharda University
Greater Noida, Uttar Pradesh, India

Tuhina Raj
Sharda University
Greater Noida, Uttar Pradesh, India

Raj Sandu
AIBI Higher Education
Sydney, New South Wales, Australia

Smitha Shivshankar
Australian International Institute Of Higher Education
Sydney, New South Wales, Australia

Prabhu Jyot Singh
CQUniversity
Sydney, New South Wales, Australia

Kitty Tripathi
Babu Banarasi Das Northern India Institute Of Technology
Lucknow, Uttar Pradesh, India

Anurag Yadav
Lovely Professional University
Jalandhar, Punjab, India

Vishal Yadav
Sharda University
Greater Noida, Uttar Pradesh, India

Chapter 1

Robotics in smart manufacturing

Vishal Yadav, Anurag Yadav, and Rashmi Priyadarshini

1.1 INTRODUCTION

In the contemporary scene of modern development, the integration of robotics' technology has become a foundation for smart manufacturing. This change in perspective uses cutting-edge innovations to make interconnected, versatile, and wise assembling environments. At the core of this transformation, lies the combination of advanced mechanics with information examination, man-made consciousness, and the Industrial Internet of Things (IIoT), introducing a period where assembling processes are computerized as well as progressively receptive to ongoing bits of knowledge [1].

The role of robotics in smart manufacturing extends beyond mere automation; it epitomizes a cooperative energy among physical and computerized frameworks, improving effectiveness, adaptability, and by and large, execution. As assembling conditions become progressively perplexing and interconnected, advanced mechanics turns into the key part, organizing consistent coordination between different parts of the creation lifecycle.

This section aims to dive into the essential role played by robotics with regards to smart manufacturing, investigating its key standards, applications, and the cooperative relationship it imparts to other state-of-the-art innovations. By looking at the combination of mechanical technology with information-driven investigation and insightful frameworks, we try to unwind the complex effect it has on present-day producing practices. Through this investigation, we lay the foundation for a thorough comprehension of how robotics' technology is reshaping assembly in the time of smart manufacturing.

1.1.1 Definition of smart manufacturing

The idea of "smart manufacturing" has become a massive need of the quickly changing world of contemporary industry, bringing about an

DOI: 10.1201/9781003537083-1

essential renovation of conventional production methods. Advanced technology, data analytics, and intelligent systems are all combined in smart manufacturing to create extremely productive, flexible, and networked manufacturing environments in smart robotics. In order to optimize the entire production process, smart manufacturing requires the seamless integration of multiple components, such as automation, robotics, artificial intelligence, and data analytics [2]. This section will examine the basic definition of smart manufacturing, as well as its main goals, guiding principles, and the overall impact it will have on the future of industrial processes and the automation process, which will be implemented. This section attempts to give readers a thorough understanding of the paradigm shift from conventional manufacturing to intelligent and interconnected systems through an analysis of the fundamental components of smart manufacturing [3]. In the following sections of this chapter, we will delve deeper into the various scenarios that go into making smart manufacturing a reality by providing a clear definition of smart manufacturing.

1.1.2 Integrating robotics in manufacturing processes

Industry is completely transformed by the integration of robotics into industrial processes, which improves accuracy, productivity, safety, efficiency, and results in large cost savings. Robotics focuses on smart analytics and service innovation to enhance operational capabilities and complexities, which is in line with Industry 4.0 aims. This shift optimizes labor costs and operational efficiency while utilizing big data to facilitate better decision-making capabilities [4].

One of the main advantages of robotics is that it reduces dangers for human workers by automating dangerous activities, which improves workplace safety. Furthermore, robots' repeatability and precision improve the product quality by promising consistent results in activities requiring high precision [5]. Due to their unparalleled flexibility and adaptability, robots allow manufacturers to quickly adjust to changing production needs. To meet market demands and maintain efficiency, this ability is essential. Robotic systems, which incorporate sensors and data-gathering techniques, enable the production of actionable insights that drive continuous improvement and process optimization [4].

Robotics offers organizations a competitive edge in the manufacturing industry through its various advantages, which includes cost reduction, consistent product quality, and enhanced productivity. An important component of preserving competitiveness on a global basis is automation. Additionally, robotics solves labor shortages by doing the physically taxing or skill-specific activities, providing a dependable workforce for complex processes [4], as mentioned in Figure 1.1.

Figure 1.1 Automatic robotics' systems. ("Image: Freepik.com".)

1.2 HISTORICAL OVERVIEW OF ROBOTICS IN MANUFACTURING

1.2.1 Evolution of robotics in manufacturing

The historical route of robotics in manufacturing uncovers an excited narration of technological progress and the transformative changes in industrial processes and automation, marked by the important milestones and innovations in history as marked below.

- **1950s – Emergence of Industrial Robots:**
  The introduction of the first industrial robots is around the 1950s, exemplified by the "Unimate" in 1961, signified the advent of automation in manufacturing. This period is patterned "The First Industrial Robot" by G. C. Devol and J. Engelberger [6].
- **1960s – Programmable Logic Controllers (PLCs):**
  The 1960s witnessed the integration of programmable logic controllers (PLCs), an important development in enhancing the capabilities of industrial robots. The impact of PLCs are discussed in "Industrial Robots: Computer Interfacing and Control" by B. J. Davies [7].
- **1970s – Advancements in Computer Technology:**
  The 1970s marked the significant progressions in computer technology, laying the foundation for the integration of microprocessors

and microcontrollers into robotics. This period is investigated in "Robot Innovation Basics" by J. D. Goddard [8].

- **1980s – Industrial Robots Become Prolific:**
The 1980s marked a period for industrial robots, with increased adoption rate across the industries. The evolution of robots during this decade is outlined as "Robotics and Automated Manufacturing" by H. Asada and K. Kosuge [9].
- **1990s – Rise of Collaborative Robots:**
The last part of the 1990s saw the development of cooperative robots, changing the scene of assembling. This shift is examined in "Cooperative Mechanical technology: An Extensive Audit" by S. Khatib [10].
- **2000s – Integration of Vision Systems:**
The 2000s brought up the integration of advanced vision systems into robotics' applications. The impact of vision-guided robots can be seen in detail in "Vision-Based Robot Guidance" by R. Kuc [11].
- **2010s – Industry 4.0 and Smart Manufacturing:**
The 2010s marked the era of Industry 4.0 and smart manufacturing, showcasing the central role of robotics. The growth of robotics and Industry 4.0 is explored in "The Fourth Industrial Revolution" by K. Schwab [12].
- **2020s – Continuing Advancements and AI Integration:**
This overview focuses on highlighting the historical progression of robotics in manufacturing, showcasing a rapid increase in efficiency, precision, and adaptability.

1.2.2 Impact on traditional manufacturing practices

The integration of robotic technology into manufacturing has powerfully changed customary works, introducing another time of productivity, cost-viability, and wellbeing. Computerized frameworks, famous for accuracy, have smoothed out work processes, limiting process durations, and generally enhancing yield [13]. This advancement has essentially diminished costs, helping with work expenses and limiting imperfections, and encouraging asset enhancement. Security and working environment conditions have uniquely worked on undertakings that are designated to robots, diminishing the gamble of mishaps, and guaranteeing the prosperity of human specialists [14]. In addition, advanced mechanics has raised item quality and accuracy, adding to normalized and solid results in assembling processes [15].

The adaptability presented by advanced mechanics circumvents difficulties related to adjusting the changing creation prerequisites. Quicker reconstructing and retooling allow quick reactions to shifts sought after or modifications in an item plan. Also, mechanical technology works with information-driven navigation, offering ongoing bits of knowledge that upgrade tasks and improve generally productivity [16].

The essential reception of mechanical technology positions organizations seriously in the worldwide market by further developing productivity, diminishing expenses, and reliably conveying great items [17]. In outline, the mix of advanced robotics has changed conventional assembling, delivering productivity, security, and seriousness.

1.3 KEY COMPONENTS OF ROBOTICS IN SMART MANUFACTURING

In the area of smart manufacturing, mechanical technology serves as a crucial part, driving proficiency, versatility, and development. This segment centers around the vital parts of advanced robotics, accentuating the jobs played by modern robots and cooperative robots (Cobots).

1.3.1 Robotic systems

At the center of smart manufacturing, robotic systems structure the foundation of automated processes, organizing exact and dynamic communications inside manufacturing conditions.

1.3.1.1 Industrial robots

Current robots have reformed the area of gathering automation, flaunting unmatched accuracy, and solidness. These mechanical wonders are intended for a large number of uses, from gathering and welding to material dealing with and painting. The invention of modern robots, exemplified by the spearheading of "Unimate," denoted an extraordinary period in assembling [6]. (Unimate was the first industrial robot, which worked on a General Motors assembly line in 1961. It has since made significant contributions to manufacturing and civilization, and has led to the robotics' field expanding beyond manufacturing). Their continuous development proceeds to essentially upgrade the productivity and adaptability of contemporary creation processes, driving constant improvement.

1.3.1.2 Collaborative robots (Cobots)

As opposed to their modern partners, cooperative robots, or Cobots, are intended to work with human administrators, cultivating a synergistic connection among man and machine. Cobots assume an urgent part in upgrading adaptability and security on the production line floor, permitting people and robots to team up in shared work areas. Their natural points of interaction and high-level sensor advances empower them to adjust to dynamic conditions, making them especially appropriate for unpredictable errands and little cluster creations [18]. The ascent of Cobots addresses a change in outlook towards human–robot cooperation, rethinking the conceivable outcomes of brilliant assembling.

In the resulting areas, we dig further into the qualities, applications, and progressions inside modern robots and cooperative robots, disentangling their extraordinary effect on the assembling scene.

1.3.2 Automation systems

Inside the domain of smart manufacturing, computerization frameworks assume a vital part in coordinating consistent and exact tasks. This part dives into two vital parts of robotization frameworks: programmable rationale regulators (PRRs) and PC computerized numerical control (CNC) machines, as seen in Figure 1.2.

1.3.2.1 Programmable logic controllers (PLCs)

Programmable logic controllers, regularly known as PLCs, act as the focal sensory system of assembling mechanization. These hearty electronic gadgets are modified to control and screen different modern cycles. At first created to supplant permanently set up transfer frameworks, PLCs have developed into modern regulators equipped for overseeing complex arrangements and assignments. Their adaptability and flexibility make them indispensable to the robotization of undertakings, for example, robotics' systems, automated control, and information obtaining. The use of PLCs improves the productivity, dependability, and constant responsiveness of assembling frameworks, contributing fundamentally to the worldview of smart manufacturing [19].

1.3.2.2 Computer numerical control (CNC) machines

In the domain of accuracy machining, PC mathematical control (CNC) machines stand as a foundation of robotization. These machines, driven by mathematical control frameworks, exactly execute producing cycles like processing, turning, and added substance fabricating. CNC machines make an interpretation of advanced plan particulars into actual items with unrivaled precision. Their capacity to create perplexing parts and models with consistency has upset the assembling scene. As a critical part of smart manufacturing, CNC machines work with the creation of complex and modified parts, adding to expanded effectiveness and quality in assembling processes [20].

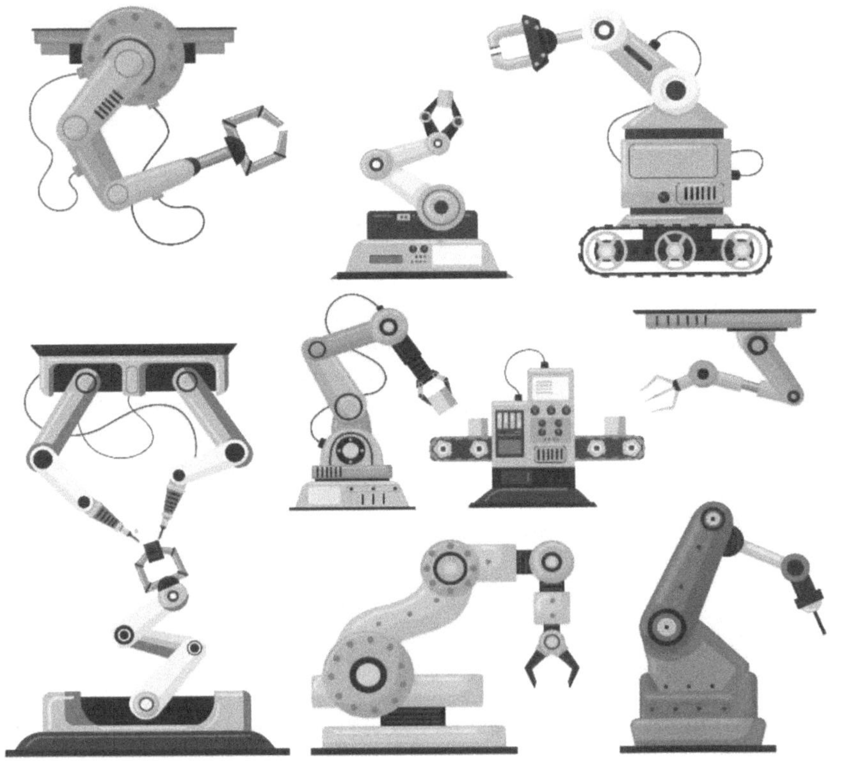

Figure 1.2 Autonomous systems. ("Cover designed by Freepik".)

1.3.3 Sensing and perception technologies

In the scene of smart assembling, detecting and discernment advancements act as the eyes and ears, empowering savvy navigation and versatility. This part investigates two urgent parts inside this area: vision frameworks and sensors and actuators.

1.3.3.1 Vision systems

Vision frameworks are instrumental in furnishing machines with the capacity to outwardly see and decipher their current circumstance. These frameworks include cameras, picture sensors, and high-level calculations to catch, process, and dissect visual data. In assembling, vision frameworks assume an essential part in quality control, object acknowledgment, and route. The reconciliation of the man-made brainpower and AI further upgrades their abilities, empowering versatile and dynamic reactions to visual improvements. Vision frameworks add to the development of smart assembling by giving continuous bits of knowledge, further developing item quality, and improving, in general, functional productivity [21].

1.3.3.2 Sensors and actuators

Sensors and actuators structure the tactile sensory system of robotized frameworks, empowering them to associate with the actual world. Sensors distinguish changes in the climate, like temperature, strain, or closeness, and convert this data into electrical signs. Actuators, then again, answer these signs by starting mechanical activities. In assembling, sensors and actuators are vital to handle checking, criticism control, and guaranteeing accuracy in different tasks. Their part in establishing a responsive and versatile assembling climate lines up with the standards of shrewd assembling, where constant information and dynamic changes add to improved proficiency and quality [22].

1.4 APPLICATIONS OF ROBOTICS IN SMART MANUFACTURING

Smart assembling use mechanical technology across assorted applications to upgrade proficiency, adaptability, and quality. This segment centers around one critical application: mechanical sequential construction systems, as mentioned in Figure 1.3.

1.4.1 Robotic assembly lines

Robotic assembly construction systems address a change in perspective in the assembling scene, where robots flawlessly team up to collect complex items with accuracy and speed. The integration of the robotics in assembly processes brings over several advantages, for example, expanded throughput, decreased process durations, and upgraded item consistency.

Case Study 1: Automotive Industry
In the automation area, robotics' construction systems have become inseparable from productivity and accuracy. Car producers use automated arms outfitted with cutting-edge vision frameworks and end-of-arm apparatuses to gather parts like case, motors, and insides. These robots work cooperatively in synchronized successions, guaranteeing exactness in multifaceted undertakings like welding, affixing, and painting. The outcome is a

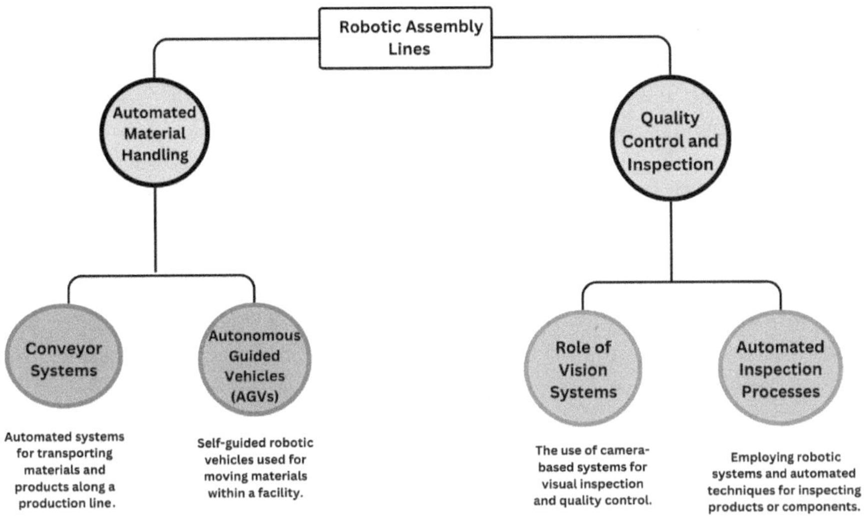

Figure 1.3 Applications of robotics in smart manufacturing.

smoothed out gathering process that upgrades creation rates while keeping up with high item quality [23].

Case Study 2: Electronics Manufacturing
In electronics' hardware fabricating, where scaled down parts request accuracy, mechanical sequential construction systems assume an urgent part. Mechanical arms with high-accuracy grippers gently handle and spot microelectronics onto circuit sheets. The adaptability of these robots takes into consideration fast reconstructing, working with the get together of different electronic items on a similar line. This flexibility is critical in the high-speed universe of hardware, where item emphases and customization are continuous [24].

These case studies represent the transformative impact of robotic assembly lines across the various industries. By automating complex and repetitive assembly tasks, manufacturers achieve keen efficiency, reduced labor costs, and improved overall product quality.

1.4.2 Automated material handling

In the domain of smart manufacturing, the utilization of mechanical technology stretches out past gathering to envelop mechanized material dealing with – a basic part enhancing planned operations and production network processes. This segment investigates two critical advancements inside this area: transport frameworks and independent directed vehicles (AGVs).

1.4.2.1 Conveyor systems

Conveyor systems form the foundation of a computerized material taking care of framework, giving a consistent and proficient method for moving materials inside assembling offices. These frameworks use a progression of mechanized belts, rollers, or chains to move merchandise along foreordained ways. In smart manufacturing, conveyor systems are coordinated with sensors and mechanization regulators to empower dynamic steering, arranging, and following. The outcome is a profoundly smoothed out material stream, decreasing manual intercession, limiting mistakes, and upgrading, in general, functional effectiveness. Conveyor systems are especially priceless in situations including nonstop creation lines, where the steady and dependable development of materials is vital for ideal and proficient assembling processes [25].

1.4.2.2 Autonomous guided vehicles (AGVs)

Autonomous guided vehicles (AGVs) address a portable mechanical answer for material vehicle inside assembling and stockroom conditions. These vehicles are furnished with sensors, cameras, and route frameworks that empower them to independently explore through spaces, staying away from deterrents and sticking to foreordained courses. AGVs assume a groundbreaking part in material taking care of by offering adaptability and versatility to changing functional prerequisites. They can be utilized, for example, for shipping natural substances, conveying parts to sequential construction systems, or moving completed items to capacity regions. AGVs add to the standards of brilliant assembling by improving the readiness and responsiveness of the material coordinated factors, at last upgrading the whole presentation biological system [26].

1.4.3 Quality control and inspection

Ensuring product quality is paramount in smart manufacturing, and robotics plays a pivotal role in automating and enhancing quality control and inspection processes. This section explores the contribution of vision systems and automated inspection processes in achieving superior quality standards.

1.4.3.1 Role of vision systems

Vision systems are instrumental in altering quality control by furnishing machines with the ability to outwardly see and decode their current circumstance. In assembling, vision systems assume a significant part in reviewing items for deserts, guaranteeing adherence to details, and keeping up with steady quality. Equipped with cameras, picture sensors, and high-level calculations, these frameworks can distinguish defects, errors, measure aspects, and confirm the rightness of get together. The constant information given by vision systems empowers fast navigation, adding to the decrease of imperfections and waste in the creation cycle. The combination of man-made consciousness further improves the capacities of vision frameworks, empowering versatile and dynamic reactions to visual stimuli, consequently raising the general quality control in assembling [27].

1.4.3.2 Automated inspection processes

Automated inspection processes influence advanced mechanics to carry out nitty gritty assessments of items with accuracy and speed. These cycles include the utilization of sensors, mechanical arms, and refined calculations to examine different parts of an item, going from surface completion to layered precision. Robots furnished with particular finish end-of-arm apparatuses can perform undertakings like nondisastrous testing, guaranteeing the underlying uprightness of parts. The computerized idea of these examination processes speeds up the appraisal as well as improves dependability by wiping out the potential for human errors. Automated inspection processes add to smart assembling by ensuring predictable quality levels, diminishing the dependence on manual examination, and working with the development of excellent items [28].

1.5 INTEGRATION OF ARTIFICIAL INTELLIGENCE IN ROBOTICS FOR SMART MANUFACTURING

In the complex texture of smart manufacturing, the combination of man-made consciousness (artificial intelligence) remains as an urgent power, introducing a time portrayed by uplifted proficiency, flexibility, and prescient capacities. This segment dives into the significant effect of AI (ML) on mechanical technology, with an engaged investigation of two basic applications: versatile control frameworks and prescient support, as mentioned in Figure 1.4.

1.5.1 Machine learning in robotics

1. Adaptive control systems

Machine earning, particularly in the domain of versatile control frameworks, arises as a key part in the development of mechanical technology for smart manufacturing. These frameworks rise above conventional control components by progressively changing boundaries in light of

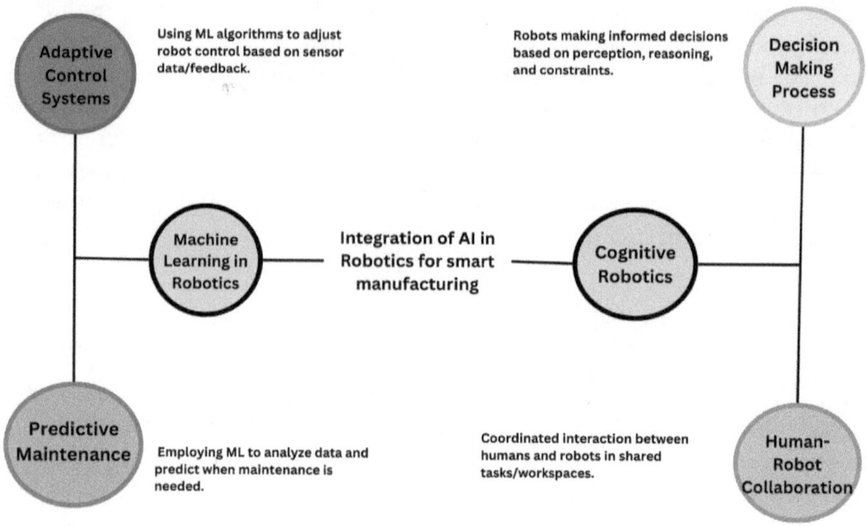

Figure 1.4 Integration of AI in robotics for smart manufacturing.

ongoing information criticism [29]. The versatile nature engages mechanical frameworks to improve their presentation in light of fluctuating ecological circumstances, guaranteeing an uplifted degree of accuracy and flexibility.

This extraordinary shift is exemplified in applications, for example, mechanical arms on sequential construction systems. Versatile control frameworks empower robots to ceaselessly refine their developments, adjusting to varieties in material attributes or natural elements [30]. The learning ability enables robots to upgrade their effectiveness over the long run, making them capable at taking care of many-sided assignments with an always working on degree of accuracy.

2. Predictive maintenance

Predictive maintenance, one more feature of AI in mechanical technology, changes the way to deal with hardware upkeep in smart manufacturing [31]. Unlike traditional maintenance practices that result in planned margin times, prescient support, driven by ML calculations, permits robots to predict possible issues by dissecting authentic information and continuous execution measurements.

By utilizing prescient upkeep, mechanical frameworks can expect part disappointments or disintegration in execution. This premonition empowers booked upkeep intercessions unequivocally while required, limiting margin time and improving the life expectancy of robotic equipment [32]. This application upgrades functional proficiency as well as adds to cost investment funds through decreased spontaneous upkeep and expanded generally speaking gear viability.

1.5.2 Cognitive robotics

In the consistently developing area of brilliant assembling, mental mechanical technology arises as a groundbreaking worldview, pervading mechanical frameworks with cutting-edge mental capacities. This segment digs into two essential elements of mental advanced mechanics: dynamic cycles and human–robot cooperation.

1. Decision-making processes
Cognitive robotics revolutionizes decision-making processes inside automated frameworks, supplying them with the capacity to dissect complex situations and pursue informed decisions. AI calculations assume a significant part in this specific circumstance, permitting robots to gain from information designs and adjust their dynamic techniques after some time [33]. This mental capacity is especially favorable in unique assembling conditions where flexibility and quick navigation are principal. One illustrative application is in independent versatile robots exploring complex production line floors. Through ceaseless learning, these robots upgrade their dynamic cycles, improving courses for effective material vehicle and adjusting to changing creation requests [34].

2. Human–robot collaboration
The cooperative energy among people and robots is reclassified through cognitive robotics, cultivating a cooperative biological system where each supplements the qualities of the other. Robots furnished with mental abilities can figure out human motions, orders, and even expect human goals [35]. This degree of cognizance works with the consistent cooperation in shared work areas, thus generally improving efficiency and safety.

An example situation is in cooperative gathering undertakings, where robots with mental abilities can decipher human guidelines and progressively change their activities to aid many-sided get-together cycles [36]. This further develops productivity as well as opens additional opportunities for human–robot organizations in assembling.

1.6 CHALLENGES AND FUTURE TRENDS

The headway of advanced mechanics in shrewd assembling delivers the two valuable open doors and difficulties. This part investigates the specialized difficulties inside this powerful scene, zeroing in on adaptability and interoperability.

1.6.1 Technical challenges

The execution of mechanical technology in smart manufacturing presents specialized difficulties that warrant creative answers for supported progress. Two key difficulties are adaptability and interoperability.

1.6.1.1 Scalability

Scalability emerges as a basic specialized challenge, particularly as assembling offices mean to grow and adjust their mechanical frameworks. The mix of extra robots or the alteration of existing frameworks to oblige changing creation volumes can introduce intricacies. Accomplishing adaptability requires arrangements that guarantee consistent coordination and correspondence between different automated parts. Tending to this challenge includes creating particular and adaptable mechanical designs that can undoubtedly scale to fulfill the advancing needs of dynamic assembling conditions [37].

1.6.1.2 Interoperability

Interoperability, the capacity of different mechanical frameworks to cooperate consistently, is a key test in shrewd assembling. With the expansion of mechanical innovations from various sellers, accomplishing a firm and interoperable environment becomes principal. Normalized correspondence conventions and points of interaction are essential for empowering interoperability, permitting different automated parts to trade information and team up really. Defeating the interoperability challenge requires broad cooperation to lay out normal guidelines that encourage a bound together and coordinated way to deal with mechanical frameworks [38]. Addressing these specialized difficulties will be fundamental for opening the maximum capacity of mechanical technology in shrewd assembling. As the business keeps on advancing, the progression of versatile and interoperable arrangements will shape the future direction of automated applications.

1.6.2 Future innovations and emerging technologies

The scene of smart manufacturing is ready for consistent advancement, driven by future developments and arising innovations. This segment investigates the thrilling possibilities that hold the possibility to reshape the field.

Innovative technologies

1. Advanced machine learning and AI algorithms: future developments in mechanical technology will probably see progressions in AI and man-made brainpower (computer-based intelligence) calculations. These advancements will improve the dynamic abilities of mechanical frameworks, empowering them to adjust and gain from constant information [39].
2. Human–robot cooperation: advancements in human–robot coordinated effort expect to make collaborations among people and robots on the manufacturing plant floor. Cooperative robots (Cobots) will advance to work close to human specialists more consistently, upgrading by and large efficiency and wellbeing [40].
3. Biomechanics and bio-inspired robotics: future advancements might draw motivation from nature, prompting the improvement of robots with biomechanical highlights. Bio-enlivened mechanical technology could alter fabricating processes by copying the productivity and versatility found in normal frameworks [41].

Emerging technologies

1. 5G connectivity: the implementation of 5G connectivity in manufacturing environments will enable faster and more reliable communication between robotic systems. This technology will facilitate real-time data exchange, enhancing the overall responsiveness and coordination of robotic operations [42].
2. Edge computing: future developments will probably use edge registering to handle information nearer to the source, lessening dormancy in automated tasks. This headway will upgrade the proficiency of dynamic cycles inside automated frameworks [43].
3. Blockchain for supply chain integration: the combination of blockchain innovation in smart manufacturing will upgrade inventory network straightforwardness and recognizability. Blockchain can be utilized to safely record and check exchanges inside the assembling biological system, cultivating trust and responsibility [44].

These future developments and arising innovations hold the possibility to drive shrewd assembling into new domains of proficiency, versatility, and maintainability. As analysts and industry experts investigate these potential outcomes, the landscape of robotics in manufacturing is ready for transformative advancements.

1.7 CASE STUDIES

1.7.1 Successful implementations of robotics in smart manufacturing

Automotive manufacturing

Cooperative advanced mechanics change sequential construction systems. In the powerful scene of car production, a convincing contextual investigation sticks out, exhibiting the consistent joining of cooperative mechanical technology into sequential construction systems [45]. This creative methodology includes the usage of cooperative robots (Cobots) working one next to the other with human administrators, aggregately arranging complicated undertakings going from appending parts to welding and painting processes. The meaning of this execution lies in the groundbreaking effect it has on the conventional sequential construction system worldview.

Collaborative robotic capabilities

These Cobots are not simple machines; they are furnished with cutting-edge sensors and vision frameworks that engage them to explore and adjust to the dynamic and developing nature of gathering undertakings. The joint effort between human administrators and Cobots presents a change in outlook, upgrading the general proficiency of the gathering system. The high-level actual abilities allow the Cobots to answer progressively to changes in the climate, guaranteeing versatility and accuracy in every phase of assembly.

Enhanced efficiency and precision

The cooperative nature of these robots brings about a significant decrease in process durations and a critical improvement in general effectiveness. By imparting the responsibility to human partners, Cobots smooth out the gathering system, guaranteeing that each undertaking is executed with the greatest possible level of accuracy. Whether it is connecting unpredictable parts or executing fragile welding and painting errands, the blend of human aptitude and mechanical precision lifts the quality and consistency of the sequential construction system yield.

Synergy between human skills and robotic precision

This contextual analysis highlights the effective cooperative energy between the nuanced abilities of human administrators and the accuracy of automated capacities. Rather than supplanting human specialists, Cobots supplement their abilities, establishing an amicable workplace. The combination of cooperative advanced mechanics rethinks the jobs of human specialists, permitting them to zero in on assignments that require imagination, critical thinking, and navigation, while Cobots handle dreary and accuracy situated exercises.

Future implications and advancements

The successful integration of cooperative mechanical technology in car fabrication makes way for future progressions in smart manufacturing. This contextual investigation fills in as a benchmark, representing how the cooperative methodology can prompt expanded efficiency, further developed working environment security, and the agreeable concurrence of human and mechanical capacities.

1.8 CONCLUSION

1.8.1 Recap of key points

Our investigation into the mix of mechanical technology in brilliant assembling has crossed a scene characterized by development, effectiveness, and groundbreaking potential. This extensive excursion started by drenching ourselves in the idea of brilliant assembling – a development that blends trend-setting innovations, information examination, and keen frameworks to develop profoundly productive, versatile, and interconnected assembling conditions.

As explored throughout this chapter, the focus moved consistently to how advanced mechanics can reshape customary assembling processes. We analyzed how mechanical technology adds to accuracy, effectiveness, cost enhancement, security, and the general improvement of item quality across diverse industrial domains.

The historical overview of robotics in manufacturing unfolded as a testament to the evolution of these technologies and their profound impact on conventional manufacturing practices. We delved into the transformative effect robotics has had on traditional methods, emphasizing its role in revolutionizing efficiency, flexibility, and adaptability in production systems.

A closer look at the key components driving smart manufacturing unfolded, incorporating mechanical frameworks, robotization instruments, and detecting innovations. This investigation is meant to give an all-encompassing comprehension of the basic components forming the area of present-day fabrication.

Our investigation extended further into the multitude of advanced mechanics' uses in brilliant assembling. From mechanical sequential construction systems to robotized material dealing with and quality control, we saw the flexibility and versatility of these advances, improving assembling processes across different areas. In addressing difficulties, we explored specialized obstacles, such as adaptability and interoperability, while additionally thinking about the moral, social, and cultural ramifications. We additionally saw into the future, investigating patterns like high-level AI,

5G network, and blockchain joining, as impetuses for proceeding with the development of smart manufacturing.

An essential part of our investigation was the assessment of true cases in auto and hardware fabricating. These contextual analyzes showed how cooperative mechanical technology alters sequential construction systems and improves effectiveness in scaled-down gathering processes, giving unmistakable instances of the extraordinary force of mechanical technology. All in all, this part has divulged the multi-layered scene of coordinating advanced mechanics in brilliant assembling. According to verifiable points of view to current applications and future patterns, the story highlights the extraordinary force of mechanical technology in molding the fate of modern cycles. As we end this section, the stage is set to proceed with investigations into the developing domains of shrewd assembling, where the cooperative energy between human resourcefulness and mechanical accuracy gets ready for another period in modern advancement. Subsequent chapters will additionally dive into arising advances, present contextual analyses, and give experiences into the continuous progressions in the unique field of mechanical technology and smart manufacturing.

REFERENCES

[1] Tao, F., Cheng, J., Qi Q., Zhang, M., Zhang, H., and Sui, F., "Digital twin-driven product design, manufacturing and service with big data," *The International Journal of Advanced Manufacturing Technology*, vol. 94, no. 9-12, pp. 3563–3576, 2018.

[2] Lee, L., Kao, H. A., and Yang, S., "Service innovation and smart analytics for Industry 4.0 and big data environment," *Procedia CIRP*, vol. 16, pp. 3–8, 2014.

[3] Wang, J., Wan, Z., and Yuan, Q., "Smart manufacturing: From conventional to intelligent and interconnected systems," *Journal of Manufacturing Science and Engineering*, vol. 142, no. 6, pp. 738–757, 2020.

[4] S. Lee, H. Kao, and G. Yang, "The role of robotics in achieving Industry 4.0 goals: Service innovation and smart analytics," *Journal of Robotics and Manufacturing*, 2014.

[5] S. Chien, T. Ma, and X. Li, "Industrial robot safety and protection systems," *Safety Science*, 2018.

[6] G. C. Devol and J. Engelberger, "The first industrial robot," *Robotics Age*, vol. 3, no. 9, pp. 10–14, 1981.

[7] B. J. Davies, *"Industrial Robots: Computer Interfacing and Control,"* Springer, 1986.

[8] J. D. Goddard, *"Robot Technology Fundamentals,"* CRC Press, 1990.

[9] H. Asada and K. Kosuge, *"Robotics and Automated Manufacturing,"* Wiley, 1985.

[10] S. Khatib, "Collaborative robotics: A comprehensive review," *IEEE Robotics and Automation Magazine,* vol. 22, no. 4, pp. 80–89, Dec. 2015.

[11] R. Kuc, "*Vision-Based Robot Guidance,*" Springer, 2003.
[12] K. Schwab, "The Fourth Industrial Revolution," *World Economic Forum*, 2016.
[13] J. Lee, H. A. Kao, and S. Yang, "Service innovation and smart analytics for Industry 4.0 and big data environment," *Procedia CIRP*, vol. 16, pp. 3–8, 2014.
[14] T. F. Chien, C. C. Ma, and H. L. Li, "Industrial robot safety and protection system: A review," *Robotics and Computer-Integrated Manufacturing*, vol. 49, pp. 212–224, 2018.
[15] L. G. Barajas and J. P. Desai, "Robotic manipulator control for high-precision tasks in manufacturing," *Robotics*, vol. 8, no. 2, p. 42, 2019.
[16] F. Tao et al., "Digital twin-driven product design, manufacturing and service with big data," *The International Journal of Advanced Manufacturing Technology*, vol. 94, no. 9-12, pp. 3563–3576, 2018.
[17] M. E. Porter and J. E. Heppelmann, "How smart, connected products are transforming competition," *Harvard Business Review*, vol. 92, no. 11, pp. 64–88, 2014.
[18] S. Khatib, "Collaborative robotics: A comprehensive review," *IEEE Robotics and Automation Magazine,* vol. 22, no. 4, pp. 80–89, Dec. 2015.
[19] T. A. Hughes, "*Programmable Logic Controllers: A Practical Approach to IEC 61131-3 Using CoDeSys,*" Elsevier, 2015.
[20] G. E. Tolvanen, "Computer numerical control of machine tools," *IEEE Transactions on Industrial Electronics,* vol. 23, no. 6, pp. 337–343, Dec. 1976.
[21] R. Klette, *"Concise Computer Vision,"* Springer, 2014.
[22] A. Rogers and R. Adams, *"Mathematical Elements for Computer Graphics,"* McGraw-Hill Education, 1976.
[23] M. Quigley, D. Gerkey, and R. Ng, *"Programming Robots with ROS: A Practical Introduction to the Robot Operating System,"* O'Reilly Media, 2015.
[24] P. Khandelwal, *"Robotics Process Automation in Electronics Manufacturing,"* IntechOpen, 2019.
[25] M. P. Groover, *"Automation, Production Systems, and Computer-Integrated Manufacturing,*" Pearson, 2018.
[26] R. M. Dolan, "Autonomous guided vehicles: A new approach to materials handling," *Material Handling Engineering*, vol. 51, no. 5, pp. 56–60, 1996.
[27] R. Klette, *"Concise Computer Vision,"* Springer, 2014.
[28] M. P. Groover, *"Automation, Production Systems, and Computer-Integrated Manufacturing,"* Pearson, 2018.
[29] S. Boyd and C. Barratt, "Linear controller design: Limits of performance via convex optimization," *IEEE Transactions on Automatic Control*, vol. 49, no. 7, pp. 798–811, 2004.
[30] R. W. Brockett, "*Finite Dimensional Linear Systems*," John Wiley & Sons, 2003.
[31] J. Lee, P. Davari, and J. W. Sutherland, "Predictive maintenance decisions for manufacturing systems considering imperfect maintenance actions," *Journal of Manufacturing Systems*, vol. 34, pp. 42–50, 2015.

[32] J. Wang et al., "A review on data-driven approaches for prognostics and health management in manufacturing," *Journal of Manufacturing Systems*, vol. 48, pp. 140–152, 2018.

[33] R. S. Sutton, A. G. Barto, "*Reinforcement Learning: An Introduction*," MIT Press, 2018.

[34] M. Quigley et al., "ROS: An open-source robot operating system," *ICRA Workshop on Open Source Software*, 2009.

[35] S. S. Srinivasa et al., "Herb: A home exploring robotic butler," *Proceedings of the Fifth International Conference on Robot Systems Integration (RSI)*, 2009.

[36] A. Sahai et al., "Human-robot collaborative assembly using a robot arm with adaptive impendence control," *Robotics and Computer-Integrated Manufacturing*, vol. 53, pp. 101703, 2018.

[37] Y. Lu, X. Xu, and Z. Zuo, "Scalability challenges and opportunities in industrial Internet of Things," *IEEE Transactions on Industrial Informatics*, vol. 14, no. 8, pp. 3489–3496, 2018.

[38] T. R. Henderson et al., "Interoperability in the industrial Internet of Things: Evaluation of DDS and MQTT," *IEEE Access*, vol. 5, pp. 6932–6939, 2017.

[39] I. Goodfellow, Y. Bengio, and A. Courville, "*Deep Learning*," MIT Press, 2016.

[40] S. Khatib, "A unified approach for motion and force control of robot manipulators: The operational space formulation," *IEEE Journal of Robotics and Automation*, vol. 3, no. 1, pp. 43–53, 1987.

[41] R. Full and D. Koditschek, "Templates and anchors: Neuromechanical hypotheses of legged locomotion on land," *Journal of Experimental Biology*, vol. 202, no. 23, pp. 3325–3332, 1999.

[42] M. Chen et al., "5G-enabled tactile internet with edge computing for control of remote systems," *IEEE Journal on Selected Areas in Communications*, vol. 34, no. 3, pp. 460–473, 2016.

[43] M. Satyanarayanan et al., "Edge analytics in the Internet of Things," *IEEE Pervasive Computing*, vol. 14, no. 2, pp. 24–31, 2015.

[44] C. Wang et al., "Blockchain-based supply chain finance: A decentralized approach," *IEEE Transactions on Engineering Management*, vol. 68, no. 3, pp. 984–996, 2021.

[45] F. Haddadin et al., "The role of the robot ankle in compensating landing height and forward momentum in the absence of anticipatory and feed-forward strategies," *IEEE Transactions on Robotics*, vol. 28, no. 4, pp. 972–982, 2012.

Chapter 2

Role of AI in smart robotic systems

Anurag Yadav, Vishal Yadav, and Rashmi Priyadarshini

2.1 INTRODUCTION

The development of artificial intelligence (AI) and robotics spans back to their common beginning during the 1950s, where the conceptualization of intelligent machines normally cleared the path for automated applications. At first, the limits between these two disciplines were obscured, given the fundamental association between the possibility of knowledge in the machines. While recognizing that not all machines qualify as robots, it is clear that artificial intelligence's degree stretches out past actual specialists to incorporate virtual elements too. Alternatively, robotics faces remarkable specialized difficulties in acknowledging actual independence, different from the worries of customary artificial intelligence research.

A notable divergence between the two fields arose during the 1970s, with robotics gradually focusing on industrial automation, while artificial intelligence expected to show machine capacities in varied real-world contexts. Regardless of enthusiasm, the mission for robots equipped for working in unstructured conditions presented huge difficulties, provoking computer-based intelligence scientists to investigate elective testbeds. In any case, this period saw significant progressions in modern robotics, driven by robotics research.

The scene moved during the 1990s, set apart by a resurgence of interest in robotics inside simulated intelligence research facilities and a renewed focus on real-world applications. This resurgence was worked with by the development of robot contests, which not just reignited the advantageous connection among simulated intelligence and robotics yet in addition prodded advancement in the two fields [1].

Today, the depiction among artificial intelligence and robotics stays tricky, yet the common objective of building insightful robots highlights a bound together exploration plan. The challenges inherent in this pursuit are very much recognized within the research community, positioning the development of robots as AI endeavors.

DOI: 10.1201/9781003537083-2

2.1.1 What is AI (artificial intelligence)?

Artificial intelligence (AI) constitutes the development of computer systems blessed with the ability to execute undertakings customarily held for human insight. These tasks span a spectrum, encompassing learning, reasoning, problem-solving, natural language comprehension, and perception. The general goal of man-made intelligence is to cause machines equipped for imitating human mental offices, subsequently empowering them to adjust, gather bits of knowledge as a matter of fact, and pursue free choices. In contemporary talk, the term AI exemplifies the complete conceptualization of machine knowledge, enveloping both functional functionalities and more societal implications.

A broadly embraced definition, as communicated by Russell and Norvig, describes artificial intelligence as "the study of human intelligence and actions replicated artificially, such that the resultant bears to its design a reasonable level of rationality" [2]. This definition highlights the double push of artificial intelligence research: understanding human comprehension and behavior while endeavoring to instantiate these principles in artificial systems.

Moreover, artificial intelligence reaches out past the domain of conventional figuring hardware, as its standards might possibly be applied to any appearance of knowledge, including natural life forms. Besides, the origination of knowledge itself envelops a rich embroidery of capacities, reaching out past coherent thinking to envelop features like imagination, the capacity to understand people on a deeper level, and mindfulness, clear in different species, including primates and other exceptional animals [3].

2.1.2 Evolution of AI and robots

2.1.2.1 Artificial intelligence

Artificial intelligence is tied in with making machines smart, so they can do things that normally need human minds, such as grasping, learning, and critical thinking. Imagine robots that can think and act like humans – that is the objective of AI.

The possibility of artificial intelligence began with the thought that computers, by taking care of images and information, could become as clever as people. This was first advanced by John McCarthy in 1956, and it was set up for artificial intelligence research.

During the 1940s and 1950s, when advanced computers arose, individuals started composing programs for them to do things like playing chess and taking care of numerical statements. As PCs got all the more impressive, particularly during the 1960s, analysts began investigating better approaches to address issues and track down arrangements. By the 1980s, we had master frameworks, which resembled computerized specialists in unambiguous fields. One of the defining moments in man-made intelligence history was the point at which IBM's Dark BLUE PC beat the world chess champion in 1997, showing how far artificial intelligence had come in strategic decision-making.

Programming language plays an important part in artificial intelligence improvement. There is LISP, which was really important in the early days because it is so flexible. Then there is PROLOG, a rational programming language that is perfect for things like regular language handling. Python showed up in the early 1990s and turned out to be really well known for man-made intelligence, particularly in language undertakings. JAVA is known for being versatile and compact, making it ideal for huge AI projects. Also, C++ is about speed, which is urgent for artificial intelligence undertakings that need fast handling. In this way, the dialects we use for simulated intelligence have developed close to the actual field, helping AI intelligence adjust and prevail in heaps of various regions [4].

2.1.2.2 Robotics

The idea of making creatures or gadgets to handle laborious tasks and lighten human burdens has roots tracing back to ancient civilizations. Since forever ago, innovative personalities have made automata, human-like gadgets pointed toward motorizing tedious errands.

The expression "robot" has a generally present-day beginning, starting from the Czech word "robota," which means "heavy work" or "forced labor." Coined by Czech writer Karel Čapek in 1920, it first referred to humanoid gadgets. In the meantime, "robotics" made its presentation in Isaac Asimov's 1942 book "*Runaround*," where he presented the well-known three laws of robotics illustrating standards for robot conduct and human cooperation [5]. Although historical literature on robotics is somewhat limited, certain works, for example, those by Ceccarelli and Gasparotto, offer important authentic viewpoints traversing old times to the modern industries [6]. The development of modern robotics, as far as we might be concerned today, started to come to fruition during the 1950s, corresponding with the rise of early types of computerizations during the modern transformation. This paper expects to dig into key achievements throughout the entire existence of modern robotics from its commencement during the 1950s through the end of the 20th century.

Modern robots' development is normally sorted into four ages. The initial three, spreading over from the 1950s to the 1990s, mark huge headways in automated capacities. The fourth era, portrayed by undeniable level clever elements like high-level calculations, legitimate thinking, profound learning, complex techniques, and cooperative way of behaving, ranges from 2000 to the present however falls outside the extent of this paper. While logical writing on the historical backdrop of modern robots remains fairly scant, certain books, for example, "*Fundamentals of Mechanics of Robotic Manipulation*" by Marco Ceccareli, alongside inward reports, give significant bits of knowledge, offering representations and conversations on this fascinating point [6].

2.1.3 Significance of AI in robotics

Computer-based intelligence (man-made reasoning) assumes an essential part in improving the capacities of robots, making them more keen, versatile, and flexible. Here is the reason simulated intelligence is so significant in advanced mechanics (see Figure 2.1):

Adaptive Learning: AI intelligence empowers robots to persistently work on their exhibition by examining information and adjusting to new data through AI calculations. This ability is especially important in powerful or eccentric conditions where robots need to change their conduct progressively [7].

Decision-Making and Independence: simulated intelligence furnishes robots with the capacity to pursue canny choices in view of perplexing information examination. By investigating different variables, robots can evaluate circumstances, focus on undertakings,

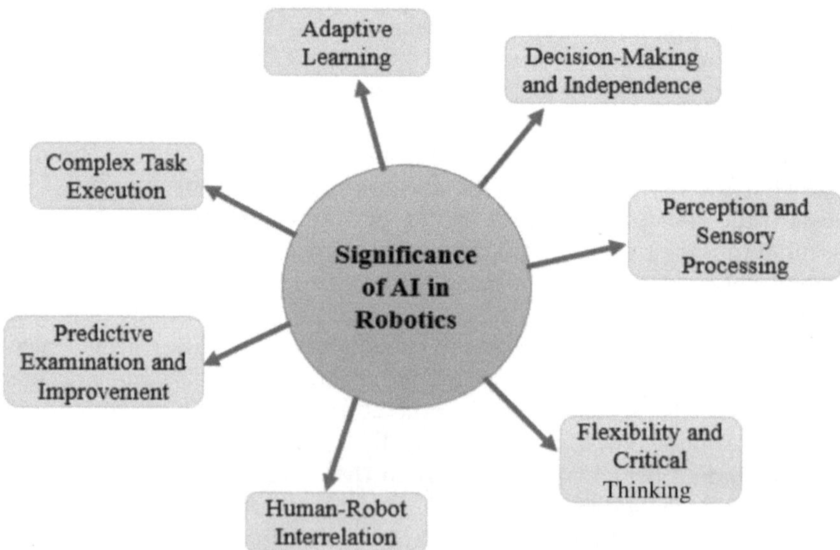

Figure 2.1 Significance of AI in robotics.

and progressively change activities, upgrading their independence and dynamic abilities [8].

Perception and Sensory Processing: AI intelligence upgrades automated discernment by handling data from different sensors. This permits robots to decipher their environmental elements, perceive articles, and make responsive moves, which is fundamental for undertakings like route and obstruction evasion [9].

Flexibility and Critical Thinking: simulated intelligence empowers robots to deal with a great many errands by applying general standards learned continuously. This flexibility makes simulated intelligence-driven robots adaptable and versatile to changing conditions and undertakings [10].

Human–Robot Interrelation: advances in normal language handling and AI work with regular associations among people and robots. Man-made intelligence empowers powerful correspondence and joint effort among people and robots, especially in settings like medical care where clear correspondence is pivotal [11].

Predictive Examination and Improvement: artificial intelligence empowers prescient examination by distinguishing examples and patterns in verifiable information. In robotics, this ability is fundamental for prescient support, course arranging, energy proficiency, and asset designation, prompting upgraded execution [12].

Complex Task Execution: AI intelligence-driven robots succeed in endeavoring complex assignments including different advances, choice focuses, and cooperations. Models incorporate complex assembling processes or fragile surgeries where exact execution is basic [13].

In summary, the meaning of artificial intelligence in robotics lies in raising frameworks past simple robotization. By empowering robots to learn, adjust, pursue shrewd choices, see and cooperate with their current circumstance, take care of mind boggling issues, and upgrade execution, artificial intelligence changes robots into flexible, wise teammates in different ventures, driving groundbreaking headways across areas.

2.2 FOUNDATIONS OF AI IN ROBOTICS

2.2.1 Relationship of AI with robotics

The connection between artificial intelligence and robotics is primary to figuring out the collaboration between these two fields. At its center, artificial intelligence gives the scholarly system to saturating robots with mental capacities, empowering them to see, reason, learn, and connect astutely with their current circumstance.

In robotics, artificial intelligence serves as the engine that drives development and progress. It empowers robots to go past straightforward mechanization and perform undertakings that require human-like insight. Through simulated intelligence strategies, for example, AI, robots can adjust to new circumstances, gain as a matter of fact, and pursue choices independently. This flexibility is essential for robots to work actually in unique and unusual conditions [2].

Moreover, AI intelligence improves the perceptual abilities of robots, permitting them to decipher tangible information from the climate and get a handle on it. This incorporates undertakings like item acknowledgment, route, and obstruction evasion, which are fundamental for robots to work securely and proficiently in true settings [9].

Additionally, AI intelligence works with normal and natural association among people and robots. Progresses in normal language handling and human–robot cooperation methods empower consistent correspondence and joint effort, making robots more available and easier to understand in different applications [11].

Fundamentally, the connection among artificial intelligence and robotics is cooperative. Computer-based intelligence gives the hypothetical establishments and reasonable devices for planning smart mechanical frameworks, while robotics fills in as a testbed for propelling man-made intelligence research and applying simulated intelligence methods to genuine issues. Together, artificial intelligence and robotics drive development and push the limits of what is conceivable in computerization, independence, and keen frameworks.

2.2.2 Integration with a robotics' system

In the domain of artificial intelligence style advanced mechanics, the consistent association among discernment and actual activity is vital. This joining flourishes inside a thorough equipment and programming framework, exemplifying a wise framework with learning and thinking capacities, perceptual frameworks (contact, power, vision), and actual activity frameworks, ordinarily addressed by robot arms and end-effectors [9]. While individual progressions in these parts are essential, their actual potential unfurls when complicatedly coordinated into a durable framework.

Adaptability arises as a basic subject intently attached to mix in cutting-edge robotics. In addition to the fact that flexibility is a critical result of joining, however it is likewise a fundamental essential. The linkage dialects between subsystems ought to be over-designed to permit significant extendibility, guaranteeing that recently discovered functionalities through combination can be successfully taken advantage of. This adaptability is especially critical in modern settings where regular robots work in unsurprising conditions. As enterprises progress towards adaptable assembling frameworks, the joining of versatile robotics becomes basic to deal with uncontrolled occasions successfully, keeping away from the significant expenses related with keeping up with profoundly controlled conditions [14].

Various worldwide drives highlight the rising significance of cutting-edge advanced mechanics, looking to overcome any barrier between scholastic exploration and pragmatic modern applications. Undertakings like the Japanese "ATRA," the French "Slam" program, and the UK's proposed Progressed Robotics Venture (ARP) epitomize a cooperative methodology including scholarly and modern consortia. These drivers plan to take lab examinations to model showing, tending to true functional difficulties [15].

2.2.2.1 *Case study: Freddy 3: advancing AI-driven robotics*

Freddy 3, a development of prior projects tracing all the way back to 1969, addresses a notable man-made intelligence-based robotics' drive. This task means to make an adaptable and effective robot programming climate, integrating programmed arranging, simulated intelligence-driven tactile handling, inductive programming, robot collaboration, and flexible correspondence between robots. Freddy 3's equipment arrangement incorporates Unimate Jaguar 260 and Rhino robots, alongside VAX 11/750 and Sun workstations. The Unix-based programming climate uses Quintus Prolog, RuleMaster, C and VAL II dialects, and Receptacle EYE for 2D vision. The cooperative energy among equipment and programming, delineated in Figure 2.2, highlights Freddy 3 as a state-of-the-art try in the unique domain of artificial-intelligence-driven robotics [16]. In rundown, the joining of simulated intelligence with robotics' frameworks presents an extraordinary worldview that stretches out past the secluded progressions in discernment and actual activity. Adaptability and versatility arise as key components, especially in modern settings, where the shift towards adaptable assembling requests coordinated arrangements. The worldwide drives and the Freddy 3 contextual analysis highlight the significance of cooperative endeavors in propelling the field, displaying the pragmatic execution and capability of coordinated artificial intelligence style robotics [15].

2.2.3 Key AI technologies

2.2.3.1 *Machine learning*

Machine Learning is the science (and art) of programming computers so they can gain from information. Here is a somewhat more broad definition: "Machine Learning is the field of study that gives computers the ability to learn without being explicitly programmed." —Arthur Samuel, 1959 [17]

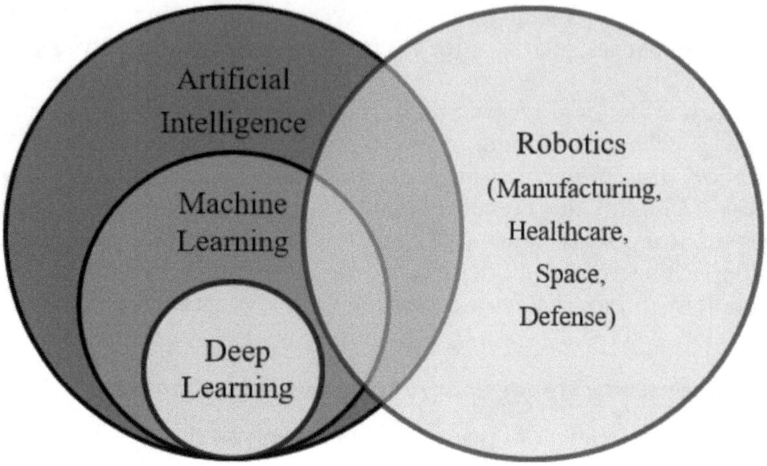

Figure 2.2 Relationship of AI with robotics.

And a more engineering-oriented one:

> "A computer program is said to learn from experience E with respect to some task T and some performance measure P, if its performance on T, as measured by P, improves with experience E." —Tom Mitchell, 1997 [18]

Machine learning, in basic terms, changes information into significant experiences. Envision it as the gathering reason behind software engineering, designing, and measurements, making it a go-to device in different fields like legislative issues and geosciences. It resembles a general critical-thinking partner for any field that arranges information understanding.

Presently, coolly, it depends on insights, and to sit back and relax; it is not quite as convoluted as it sounds. You realize how in designing; we typically take care of issues with fixed arrangements? Indeed, there are situations where we do not have the real factors or the figuring ability to make sure about an exact arrangement. That is where measurements prove to be useful.

Take figuring out human inspiration, for instance – very interesting to nail down. In sociologies, in the event that we can foresee the human way of behaving accurately 60% of the time, that is a success! Why? Since certain things are simply excessively complex to demonstrate impeccably, such as characterizing satisfaction for everyone. In this way, in these circumstances, we go to our dependable measurements' tool stash to figure out the vulnerabilities.

2.2.3.2 *Deep learning*

Diving into the universe of profound learning uncovers a change in outlook in AI. Dissimilar to its ancestors, profound learning embraces a significant way to deal with information obtaining, stripping back layers of information to reveal significant experiences. This figurative excursion looks like the fastidious unwinding of an onion, where each layer, named "representation," adds to a more profound comprehension of the basic examples.

At the center of profound learning lies brain organizations, roused by the mind boggling activities of the human cerebrum. These organizations, suggestive of engineers of data, fastidiously stack layers to remove complicated examples and connections from tremendous datasets. Nonetheless, it is basic to take note of the fact that profound learning is not an endeavor to duplicate the mind's intricacy; rather, its capabilities as a refined numerical device are proficient at unraveling complex datasets.

The advancement of profound learning has made it ready for elective descriptors, for example, "layered portrayals' learning," or "progressive portrayals' learning." These terms epitomize the pith of profound learning's various leveled structure, where the profundity of understanding increments with the quantity of layers navigated [19].

In rundown, profound learning rises above conventional AI approaches by embracing a progressive and significant comprehension of information. With brain networks as its spine, profound learning opens the potential for translating complex datasets and removing significant experiences [7].

2.2.3.3 *Computer vision*

Computer vision remains as a foundation in the domain of man-made brainpower, engaging machines with the capacity to decipher and comprehend visual data, similar to the human visual framework [20]. At its center, PC vision plans to overcome any issues between the advanced and actual universes by empowering machines to remove significant experiences from visual information.

The excursion of computer vision starts with the securing of crude visual information through different sensors, like cameras or LiDAR frameworks. When caught, this information goes through a progression of handling moves toward extricating important highlights and examples. These elements act as building blocks for more significant level undertakings, including object acknowledgment, picture characterization, and scene gathering.

Key to the success of computer vision calculations is their capacity to gain from information. AI procedures, especially profound learning, have altered the field by empowering models to consequently gain complex examples and portrayals from enormous scope datasets. This information-driven approach has prompted critical headways in regions like picture division, object identification, and picture age.

Besides, the interdisciplinary idea of PC vision draws from different fields like software engineering, science, and neuroscience. Strategies enlivened by organic vision frameworks, for example, convolutional brain organizations (CNNs), have demonstrated to be especially compelling in copying human visual handling and accomplishing cutting-edge execution in different PC vision undertakings.

As of late, PC vision has tracked down broad applications across ventures, including independent vehicles, medical care, observation, and increased reality (AR) [21]. These applications exhibit the groundbreaking capability of PC vision in upgrading computerization, further developing navigation, and empowering new types of human–machine association.

In conclusion, computer vision addresses a combination of innovation and discernment, enabling machines with the capacity to see, decipher, and grasp the visual world. With headways in AI and interdisciplinary examination, PC vision keeps on pushing the limits of what is conceivable, opening new open doors for development and revelation.

2.3 AI-DRIVEN REASONING AND DECISION MAKING

2.3.1 Why should a robot use AI-type reasoning?

To explore the intricacies of genuine errands, robots should have the capacity to reason independently utilizing advanced artificial intelligence methods. This is fundamental when errands cannot be comprehensively predefined, expecting robots to derive fitting activities in light of dubious or fragmented data. Thinking in this setting includes utilizing a range of computer-based intelligence techniques, including information portrayal, heuristic pursuit, and AI [22].

Consider the errand of a robot getting an item from a table. The sheer number of choices included – for example, which hand to utilize, how much power to apply, and where to put the gripper – makes unequivocal programming unrealistic. Besides, the setting in which the assignment is performed extraordinarily impacts direction. For example, the technique for getting a glass of juice varies from that of a delicate thing or a jumbled scene. In such situations, artificial intelligence thinking techniques become fundamental, permitting robots to gather proper activities in view of logical prompts.

Additionally, executing complex assignments frequently includes groupings of activities and cooperations with the climate. Prescient thinking empowers robots to expect the results of planned activities, while analytic thinking recognizes the reasons for deviations from anticipated conduct.

Imagining investigates conceivable occasion successions, while inquiry responding to includes deriving missing information to satisfy conditions for task execution.

To address these difficulties, independent robots utilize thinking strategies to address and reason about their abilities, surroundings, articles, and activities. These envelop a range of thinking errands, including expectation, finding, and inquiry responding to, empowering robots to explore mind boggling and dubious conditions independently [23].

All in all, artificial-intelligence-driven thinking is pivotal for engaging robots to deal with certifiable undertakings successfully, empowering them to pursue independent choices and adjust to dynamic conditions.

2.3.2 Robotic systems with automation

2.3.2.1 Role of automation in robotics

Computerization is vital in the domain of robotics, effectively smoothing out processes, reducing human exertion, improving proficiency, and increasing efficiency. It incorporates the usage of machines, including robots, to execute assignments with negligible or no human intercession. From basic dreary activities to perplexing work processes, computerization in robotics traverses a wide range of utilizations.

2.3.2.2 Types of automation in robotics

Task computerization: errand mechanization involves programming robots to tediously perform explicit predefined undertakings. These undertakings frequently incorporate exercises like picking and putting objects on an assembling sequential construction system, bundling items, or leading quality examinations. Via mechanizing such assignments, robots work on tedious tasks, decrease mistakes, and speed up creation speed [24].

Process automation: dissimilar to task computerization, process robotization spins around computerizing whole work processes or cycles as opposed to individual undertakings. This approach includes the reconciliation of different robots and frameworks to consistently coordinate complex tasks. A great representation is mechanized stockroom frameworks where robots handle stock administration, request satisfaction, and coordinate factor tasks. Process robotization upgrades proficiency, exactness, and asset use, prompting smoothed out activities and improved execution [25].

Fundamentally, mechanization fills in as a foundation in robotics, reforming businesses by empowering robots to embrace undertakings and cycles independently, in this way driving proficiency and efficiency higher than ever.

2.3.3 Autonomous navigation and path planning

2.3.3.1 Path planning workspace modeling

Way arranging is a basic part of independent routes in robotics, including the assurance of practical ways for a robot to explore from its ongoing area to a predefined objective while keeping away from obstructions and sticking to different limitations. A basic prerequisite for powerful way arranging is the accessibility of itemized data about the robot's current circumstance, generally alluded to as work area demonstrating.

2.3.3.2 Types of information in workspace modeling

Obstacle representation: one urgent part of work area displaying is the portrayal of impediments present in the climate. This data normally incorporates the area, shape, size, and properties of snags that might block the robot's development. Different strategies like occupancy grids, geometric models, or point cloud representations might be utilized to depict deterrents in the climate [26].

Surface elements: also, work area displaying may incorporate data about surface highlights that are pertinent to way arranging. This could incorporate territory height, harshness, grinding coefficients, or different properties that impact the robot's headway. Understanding surface elements permits the organizer to choose ways that streamline factors, for example, crossing speed, energy effectiveness, or strength.

2.3.3.3 Criteria for path calculation

The measures utilized for working out ways rely upon the particular prerequisites of the robot and the main job. For example, a way organizer might focus on limiting way length to effectively arrive at the objective. In such cases, the organizer principally centers around recognizing navigable regions and keeping away from deterrents while looking for the briefest way between the robot's ongoing position and the objective area.

Nonetheless, different measures might become an integral factor in view of the robot's capacities and mission targets. For instance, in the event that energy proficiency is really important, the organizer should think about the robot's energy utilization and landscape qualities. This requires integrating information on terramechanics, for example, soil type, slant, and surface circumstances, into the arrangement process. Moreover, factors like the robot's guiding component and mobility might impact way streamlining to limit energy consumption [27].

Basically, a successful way of arranging depends on thorough work area displaying and the thought of different rules custom-made to the robot's abilities and mission goals. By precisely addressing the climate and improving way choice, independent robots can explore securely and productively in different situations.

2.3.4 Example of AI-driven decision making in robotics' application

An illustrative illustration of artificial-intelligence-driven dynamics in robotics' application is the field of independent vehicles, especially self-driving vehicles. These vehicles depend on cutting-edge simulated intelligence calculations to see the climate, pursue constant choices, and explore securely to their objections.

2.3.4.1 Perception and sensing

Self-driving vehicles are outfitted with a plenty of sensors, including cameras, LiDAR (light location and going), radar, and ultrasonic sensors, to assemble data about the general climate. Man-made intelligence-driven PC vision calculations break down information from these sensors to distinguish and recognize articles, for example, walkers, vehicles, street signs, and traffic lights. Moreover, AI models are used to order protests, anticipate their developments, and evaluate possible dangers in the vehicle's way [28].

2.3.4.2 Decision making and planning

When the climate is seen, simulated intelligence calculations utilize dynamic cycles to design the vehicle's direction and explore through complex situations securely. These calculations integrate different factors, for example, transit regulations, street conditions, close by vehicles' ways of behaving, and expected snags to decide the ideal strategy. Support gaining procedures empower the vehicle to gain from previous encounters and adjust its way of behaving to various driving circumstances [29].

2.3.4.3 Example scenario

Consider a self-driving vehicle moving toward a crossing point with a traffic light going to become red. The simulated intelligence framework should choose whether to brake and stop or continue through the convergence in view of a few elements. It assesses the traffic signal status, the vehicle's ongoing pace and distance from the crossing point, the presence of different vehicles, and the probability of securely clearing the convergence before the sign changes.

2.3.4.4 AI-driven decision-making process

- **Perception:** the vehicle's sensors distinguish the traffic signal's variety and recognize the presence of vehicles and people on foot nearby.
- **Examination:** simulated intelligence calculations investigate the sensor information and evaluate the gamble related with various strategies, taking into account factors like the vehicle's force, halting distance, and encompassing traffic conditions.
- **Navigation:** in light of the examination, the artificial intelligence framework chooses whether to dial back and stop at the convergence or continue through it. This choice is affected by security contemplations, transit regulations, and the likely effect on other street clients.
- **Activity:** the vehicle executes the picked game-plan, either reaching a controlled stop at the crossing point or continuing through it warily, changing its speed and direction as it is important to stay away from crashes.

2.3.4.5 Conclusion

The model features how man-made intelligence-driven direction empowers independent vehicles to explore complex situations securely and proficiently, exhibiting the groundbreaking capability of man-made intelligence in robotics applications.

2.4 SENSOR FUSION AND PERCEPTION IN AI INTEGRATION

2.4.1 Role of AI in sensory fusion and perception

Sensory fusion and perception lie at the center of empowering robots to interface successfully with their current circumstance. This coordination is basic for robots to comprehend their environmental elements precisely, go with informed choices, and explore independently. Man-made reasoning (man-made intelligence) assumes a vital part in improving tangible combination and discernment in mechanical frameworks, adding to their knowledge and flexibility.

2.4.1.1 Sensory fusion

In current robotics, robots are outfitted with various sensors, each giving extraordinary experiences into the climate. These sensors incorporate cameras, LiDAR, radar, ultrasonic sensors, and inertial sensors, among others. In any case, individual sensors might have restrictions or

vulnerabilities in their estimations. Tangible combination, otherwise called sensor information combination, addresses this test by coordinating information from various sensors to make a complete and dependable portrayal of the climate.
Artificial intelligence procedures, especially AI calculations, are instrumental in sensor combination errands. These calculations cycle information from various sensors, make up for sensor blunders or clamor, and join data to create a bound together view of the environmental elements. For instance, profound gaining models can figure out how to extricate significant elements from sensor information and break them to further develop object identification, limitation, and planning [30].

2.4.1.2 *Perception improvement*

Perception improvement includes extricating valuable data from sensor information to comprehend the climate better. Man-made intelligence-driven discernment calculations empower robots to decipher tangible data sources, perceive objects, surmise spatial connections, and expect changes in their environmental factors. PC vision methods, controlled by profound learning, are especially successful in improving discernment by examining visual information.

Deep learning models can examine pictures or recordings caught by cameras to recognize objects, characterize them into various classes, and gauge their stances or directions. These calculations can likewise fragment pictures to recognize locales of interest and concentrate important data for direction. Furthermore, man-made intelligence calculations can handle information from different sensors, like LiDAR or radar, to supplement visual insight and make a more complete comprehension of the climate [31].

In summary, artificial intelligence assumes a pivotal part in progressing tactile combination and discernment in mechanical frameworks. By utilizing AI and PC vision strategies, robots can incorporate data from assorted sensors, make up for sensor impediments, and improve how they might interpret the climate. This joining of artificial intelligence into tactile combination and discernment adds to the knowledge and independence of automated frameworks, empowering them to perform complex assignments in true conditions.

2.5 HUMAN–ROBOT INTERACTION: AI ADVANCES

2.5.1 AI enhancements in human–robot interaction

The development of PCs has upset different parts of human existence, essentially by their capacity to perform estimations and store information at remarkable paces. With the appearance of the web during the 1980s and 1990s, the world saw another extraordinary shift. The web associated computers around the world, working with correspondence and data trade on a phenomenal scale.

One of the critical progressions in innovation that has arisen lately is the scaling down and improvement of registering gadgets. Engineers have prevailed with regards to making PC parts quicker, more modest, and more available than any time in recent memory. Noteworthy gadgets like ARDUINO and Raspberry Pi embody this pattern, offering strong registering capacities in minimized, easy to use bundles. The expansion of Wi-Fi chips has additionally worked with availability, empowering even small items to be incorporated into the web of things (IoT).

The IoT, an idea that rotates around interfacing regular items to the web, has opened up additional opportunities for human–robot connection. With the assistance of installed chips and circuits, objects going from apparatuses to vehicles can now be controlled and observed from a distance. These implanted frameworks give special characters to every gadget and empower remote correspondence, establishing the groundwork for smart, interconnected frameworks.

To change a machine into a shrewd gadget utilizing artificial intelligence, a few key cycles are involved:

- Relegating remarkable character: each machine is doled out by an interesting personality, normally through conventions like IPv6, guaranteeing that it tends to be recognized from different gadgets internationally.
- Empowering correspondence: remote correspondence abilities are fundamental for consistent connection between gadgets. Advancements like Wi-Fi empower gadgets to speak with one another and with outside frameworks.
- Joining of sensors: sensors assume an urgent part in furnishing machines with the capacity to see and cooperate with their current circumstance. Sensors, for example, pressure sensors and RFID trackers can be incorporated into machines to accumulate important data.

By incorporating these parts and utilizing man-made intelligence calculations, machines can become smart elements fit for independently collaborating with clients and their environmental factors. For instance, a seat furnished with sensors and RFID innovation can send information about inhabitants and client personality to a cell phone, empowering remote checking and control.

Fundamentally, simulated intelligence upgrades in human–robot cooperation work with the making of shrewd conditions where gadgets consistently impart and team up to improve productivity and comfort [32].

2.5.2 Presence of humanoid and biological-inspired robots in society, example Sofia

The combination of humanoid and naturally propelled robots into society denotes a critical headway in robotics, obscuring the lines among human and machine connections. One unmistakable illustration of such a robot is Sophia, created by Hanson Robotics. Sophia earned worldwide respect for her human-like appearance and capacity to participate in discussions. Sophia addresses a union of state-of-the-art robotics, computer-based intelligence, and human-like elements, making her a convincing illustration of the capability of humanoid robots in different applications. Furnished with refined facial acknowledgment innovation, discourse acknowledgment, and regular language handling capacities, Sophia can cooperate with people in a way suggestive of human-to-human correspondence.

Her presence in the public eye fills in as a demonstration of the headway made in robotics and simulated intelligence, displaying the conceivable outcomes of coordinating robots into different jobs, including client care, schooling, and diversion. Sophia has shown up at various occasions around the world, including gatherings, TV programs, and meetings, where she participates in conversations on different points, exhibiting her conversational capacities and interactive abilities.

However, it is fundamental to recognize the moral and cultural ramifications of humanoid robots like Sophia [33]. As robots become progressively similar and equipped for copying the human way of behaving, questions emerge with respect to their privileges, obligations, and effect on human culture. Discussions encompassing the moral utilization of artificial intelligence, information security, and the potential for work relocation highlight the requirement for smart thought and guidelines as humanoid robots become more common in our day-to-day routines.

In conclusion, the presence of humanoid and organically enlivened robots like Sophia features the fast advancement in robotics and simulated intelligence, offering looks into a future where human–machine collaboration is consistent and pervasive. While their joining into society presents invigorating open doors, it likewise brings up significant moral and cultural issues that require cautious consultation.

2.6 AI TRANSFORMING INDUSTRIES

2.6.1 Industry 4.0 and manufacturing

2.6.1.1 Introduction

Industry 4.0 marks the beginning of a new era in manufacturing that is marked by the integration of cutting-edge technologies. The reshaping of global industrial landscapes caused by the convergence of Industry 4.0 and artificial intelligence (AI) robotics is the subject of this section. Industry 4.0 envelops a different scope of mechanical progressions, including the Modern Web of Things (IIoT), enormous information examination, distributed computing, and digital actual frameworks [34]. These components structure the groundwork of brilliant plants and nimble creation frameworks, promising increased effectiveness and canny navigation.

2.6.1.2 Integration of AI robotics

In Industry 4.0, man-made intelligence mechanical technology reforms customary assembling ideal models by presenting versatility and knowledge. Dissimilar to traditional robots, artificial-intelligence-fueled frameworks show mental capacities, rising above routine errands and cultivating a responsive aspect in modern robotization. This integration is a paradigm shift in which industries embrace AI-driven intelligence and dynamic processes. The marriage of computerized reasoning and mechanical technology moves Industry 4.0 into a period of development, where flexibility turns into an essential basis [35].

2.6.1.3 Enhanced automation and efficiency

Simulated intelligence mechanical technology in Industry 4.0 drives elevated mechanization, empowering consistent cooperation among people and machines [36]. This collaboration improves creation plans, screens hardware wellbeing, and refines asset use, prompting upgraded proficiency and efficiency in brilliant manufacturing plants. Furnished with AI calculations, computer-based intelligence-driven robots progressively change their conduct in view of ongoing information, cultivating dexterity, and responsiveness under conditions.

2.6.1.4 Predictive maintenance and challenges

Simulated intelligence-driven mechanical technology assumes a critical part in prescient upkeep, limiting margin time and diminishing support costs by expecting potential disappointments [37]. Ongoing shortcoming identification empowers fast reactions to peculiarities, guaranteeing continuous creation processes. However, in order to fully utilize the advantages of AI robotics in Industry 4.0, careful consideration must be given to issues like data security, technology interoperability, and ethical considerations.

2.6.1.5 Conclusion

The combination of Industry 4.0 with artificial intelligence mechanical technology means a groundbreaking change in assembling, driving development, and effectiveness. This segment highlights the requirement for interdisciplinary exploration to address difficulties and open the maximum capacity of this powerful collaboration, molding the fate of modern creation on a worldwide scale.

2.6.2 Healthcare robotics: AI-assisted applications

Artificial intelligence helped applications. The combination of cutting-edge innovations, quiet man-made reasoning (artificial intelligence), and mechanical technology, has catalyzed a significant change in medical care administration experiences. Utilizing the research that Seong No Yoon and DonHee Lee conducted in their article titled "Artificial intelligence and robots in healthcare: What are the success factors for technology-based service encounters?" as a source of inspiration, this investigation delves into the changing environment in which technology plays a significant role in reshaping interactions within the healthcare sector [38].

2.6.2.1 Reconceptualizing service encounters

Service encounters are surpassing traditional human-to-human interactions in healthcare. A multifaceted framework in which humans, AI robots, and machines collaborate to provide services is introduced by the emergence of AI and robotics. This development moves the help experience structure from the customary buyer supplier dynamic to a nuanced shopper simulated intelligence robot/machine-supplier design. Increased operational efficiency, increased patient engagement, and innovative healthcare solutions are all made possible by this transition.

2.6.2.2 Roles and impacts

Inside this mechanically interceded medical care environment, particular jobs arise. As consumers, patients are encouraged to actively participate in technological advancements and embrace them, fostering a collaborative approach to healthcare delivery. By explaining the advantages and potential drawbacks of adopting technology, frontline medical staff play a crucial role in encouraging patient involvement. Simultaneously, advancements should be planned with ease of use, comfort, exactness, and monetary effectiveness at the top of the priority list. The achievement factors lattice, enveloping human (client, representative), simulated intelligence robots, and machines, fills in as a directing structure for upgrading different help experience situations.

2.6.2.3 Mechanical development in medical care

The medical services' area is seeing a multiplication of man-made intelligence helped applications, going from brilliant wearables that screen cardiovascular exercises to simulated intelligence controlled indicative instruments. The two new businesses and laid out tech organizations add to a climate where artificial intelligence robots, exemplified by IBM's Watson for Oncology, offer oncology arrangements and chance evaluations for patients. The mix of artificial intelligence upgrades demonstrative precision as well as upsets patient–supplier connections, conveying customized and productive medical care administrations.

2.6.2.4 Conclusion

As medical care embraces simulated intelligence helped applications, the achievement factors for innovation-based help experiences expect basic significance. This scholarly request lays the foundation for understanding the advancing jobs of people, artificial intelligence robots, and machines in reshaping the conveyance of medical care administrations. By investigating the complexities of innovation-based help experiences, medical services' suppliers and specialists gain important bits of knowledge into improving patient consideration, upgrading functional proficiency, and exploring the groundbreaking excursion towards an innovatively enhanced medical services' future.

2.7 ETHICAL AND FUTURE CONSIDERATIONS

2.7.1 Ethical implications in AI-driven robotics

The fast headway of man-made consciousness (computer-based intelligence) and mechanical technology delivers a large group of moral difficulties that request cautious thought. The accountability of the engineers working on these systems is a major concern. Attention to potential moral difficulties is vital to forestall abuse and guarantee straightforwardness. In the creation and implementation of AI and robotic technologies, engineers must place an emphasis on accountability by prioritizing the incorporation of mechanisms that permit human inspection of the functionality of algorithms and systems.

As independence expands in cutting-edge frameworks, the requirement for a moral dynamic inside the actual frameworks becomes basic. With the increasing number of autonomous systems working together, the risk of unintended behavior increases. A speculative situation introduced in "Moral Machines" [39] outlines the possible results of "deceptive" choices made via independent mechanical exchanging frameworks. This situation highlights the significance of framework control components, pushing for robotized impediments on conduct and systems that illuminate human administrators about conditions requiring human survey. The design and operation of these intelligent systems ought to take into account the moral implications of AI actions, such as compliance with laws and moral considerations.

Exploring the moral scene turns out to be much more many-sided while examining the consolidation of moral decision-production into man-made intelligence frameworks. Even though cutting-edge technology has a lot to offer, making decisions based on moral principles must be incorporated. The difficulty lies in making it possible for AI systems to consider legal and moral frameworks when evaluating the ethical implications of their actions. The variety and nuance of possible ethical dilemmas add complexity, necessitating a thoughtful strategy for seamlessly integrating morality into AI and robotic technologies. Engineers face the problem of offsetting mechanical progressions with moral contemplations, requiring cautious thought and the improvement of frameworks that line up with cultural qualities and standards.

2.7.2 Integration of advanced technologies

The combination of cutting-edge innovations is urgent in driving the determined movement of man-made intelligence mechanical technology, upsetting the abilities and functionalities of smart frameworks. A few extraordinary innovations are ready to lead this charge, reshaping the future scene of simulated intelligence mechanical technology:

1. **Next-Generation Connectivity (6G):** the development past 5G to 6G organizations will be instrumental in empowering continuous dynamics in advanced mechanics. With expected abilities, for example, terabit-per-second paces and close prompt reaction times, 6G organizations engage robots to flawlessly team up and work in powerful, interconnected conditions.
2. **Advanced Artificial Intelligence Models:** robots' perception, decision-making, and adaptability are improved by ongoing advancements in AI models, such as higher-capacity neural networks and advances in natural language processing. AI robotics now has unprecedented language comprehension and contextual awareness thanks to models like GPT-4 and later [40].
3. **Edge Computing and AI at the Edge:** the collaboration among man-made intelligence and edge registering diminishes inactivity and

empowers continuous reactions by handling information nearer to the source. Simulated intelligence calculations running nervous gadgets engage robots with upgraded independence, empowering effective undertaking execution without consistent dependence on incorporated servers.

4. **Quantum Computing for Complex Problem Solving:** quantum figuring holds guarantee for tackling complex advancement issues, dramatically speeding up robots' critical thinking abilities. Quantum calculations are ready to upset improvement challenges in regions like course arranging and asset portion.
5. **Advanced Sensing Technologies:** state-of-the-art detecting advances, including LiDAR, radar, and high-level cameras, upgrade robots' natural mindfulness and empower ongoing direction. Joined with simulated intelligence calculations, these sensors empower robots to explore complex environmental factors and perceive protests productively.
6. **Biologically-Inspired AI and Neuromorphic Computing:** naturally enlivened artificial intelligence and neuromorphic processing imitate brain designs, upgrading robots' learning abilities and energy proficiency. Imitating the mind's brain networks empowers robots to gain for a fact, perceive examples, and display more human-like mental capabilities [41].
7. **Robotics Operating Systems (ROS) Enhancements:** ROS continues to evolve to meet the needs of advanced AI robotics and serves as a foundation for the creation and coordination of robotic systems. Diverse robotic functionalities and components can be seamlessly integrated with ongoing enhancements [26].
8. **Augmented Reality (AR) and Mixed Reality (MR):** AR and MR advancements improve human–robot communication and coordinated effort by overlaying computerized data onto the actual world. In areas like upkeep and preparing, AR and MR assume a significant part in improving situational mindfulness and undertaking execution.
9. **Biotechnology for Biohybrid Robots:** biohybrid robots that combine biological components with robotic structures can now be made thanks to advancements in biotechnology. These biohybrid frameworks have upgraded versatility and capacities, possibly upsetting fields like medication and ecological checking.

Intelligent systems that are more autonomous, perceptive, and capable of addressing intricate real-world challenges are being developed as a result of the convergence and maturation of these technologies. Together, they propel AI robotics into new frontiers. The impact and possibilities of AI robotics in various fields are redefined by the interaction of these technological catalysts.

2.8 CONCLUSION

2.8.1 Summarize the key points discussed in the chapter

All in all, our conversation has dug into the essential job of simulated intelligence in savvy mechanical technology frameworks, as illustrated in the section. The connection between artificial intelligence and robotics, its integration with robotics systems, important AI technologies like machine learning and deep learning, and the significance of AI-driven reasoning and decision making were among the many aspects we investigated. We likewise talked about the significance of sensor combination and discernment in simulated intelligence reconciliation, close to the extraordinary effect of artificial intelligence on enterprises, medical care, and moral contemplations.

Throughout our conversation, it became clear that artificial intelligence plays a crucial role in enhancing robotics systems' capabilities. From empowering independent routes to upsetting assembling processes in Industry 4.0, man-made intelligence-driven mechanical technology offers exceptional degrees of proficiency, versatility, and efficiency. In addition, the integration of AI and robotics in healthcare holds the promise of novel approaches to the treatment and diagnosis of patients, albeit with associated ethical issues that necessitate careful consideration.

Fundamentally, the mix of man-made intelligence in brilliant advanced mechanics' frameworks means a change in perspective by the way we see and communicate with innovation. It not only gives machines the ability to carry out intricate tasks on their own, but it also requires a thoughtful approach to ethical, societal, and future considerations. As we explore this extraordinary scene, interdisciplinary coordinated effort, moral mindfulness, and mechanical development will be significant in bridging the maximum capacity of artificial intelligence in shrewd advanced mechanics' frameworks to improve society.

2.8.2 Transformative role in AI in shaping the future of smart robotics

The extraordinary job of man-made intelligence in molding the fate of shrewd mechanical technology could not possibly be more significant. As we look forward, simulated intelligence remains as the foundation of development, driving headways that reclassify the abilities and conceivable outcomes of advanced robotics' systems.

Quite possibly the main change lies in the independence and knowledge of robots. Robots can perceive and interact with their environment in real time thanks to AI algorithms, which enable them to make autonomous decisions based on intricate data analysis. This independence upgrades proficiency and efficiency as well as opens ways to new applications across different enterprises, from assembling and medical services to farming and strategies.

Moreover, simulated intelligence works with the consistent combination of robots into human-driven conditions. Cooperative robots, or cobots, furnished with simulated intelligence abilities, can work closely with people securely and proficiently, increasing human capacities and changing work processes. A new era of human–robot interaction is ushered in by this collaborative strategy, in which robots become trusted partners in the process of completing tasks with precision and dependability.

Also, artificial-intelligence-driven mechanical technology frameworks are ready to upset enterprises through prescient upkeep, customized medical services' arrangements, and versatile assembling processes. By bridging the force of simulated intelligence for prescient examination and AI, robots can expect gear disappointments, advance treatment designs, and adjust creation lines continuously, prompting cost reserve funds, further developed results, and upgraded seriousness.

In addition, artificial intelligence gives robots cognitive capabilities, allowing them to gain knowledge through experience, adjust to changing circumstances, and continuously enhance their performance over time. This capacity not just improves the adaptability and adaptability of advanced mechanics' frameworks yet in addition establishes the groundwork for the improvement of genuinely shrewd and independent machines.

In conclusion, the groundbreaking job of simulated intelligence in molding the eventual fate of brilliant advanced mechanics is obvious. By incorporating AI technologies into robotics' systems, we open up a world of possibilities in which intelligent, autonomous, and collaborative robots become essential assets for driving innovation, increasing productivity, and enhancing individual and societal quality of life. In order to realize AI's full potential in smart robotics, interdisciplinary collaboration, ethical considerations, and responsible deployment will be crucial as we embrace this future.

REFERENCES

[1] J. J. Kuffner and S. M. LaValle, RRT-Connect: An Efficient Approach to Single-Query Path Planning. In *Proceedings of the IEEE International Conference on Robotics and Automation (ICRA)* (Vol. 2, pp. 995–1001). England: IEEE, 2000.

[2] S. Russell and P. Norvig, *Artificial Intelligence: A Modern Approach*, 3rd ed. New Delhi: Prentice Hall, 2009.

[3] P. O. Haikonen, *The Cognitive Approach to Conscious Machines*. England: Imprint Academic, 2008.

[4] J. McCarthy et al., "A Proposal for the Dartmouth Summer Research Project on Artificial Intelligence," *AI Magazine*, vol. 27, no. 4, pp. 12–14, 2006. DOI: 10.5555/1610153.1610155.

[5] I. Asimov, Runaround. In *I, Robot*, New York, NY: Gnome Press, 1950, pp. 56–89.
[6] M. Ceccarelli and L. Gasparetto, *Fundamentals of Mechanics of Robotic Manipulation*, Berlin: Springer, 2013.
[7] Y. LeCun et al., "Deep Learning," *Nature*, vol. 521, no. 7553, pp. 436–444, 2015.
[8] S. Thrun et al., "Probabilistic Algorithms in Robotics," *Proceedings of the National Academy of Sciences*, vol. 95, no. 15, pp. 8420–8427, 1998.
[9] R. Siegwart and I. R. Nourbakhsh, *Introduction to Autonomous Robots*, Cambridge, MA: MIT Press, 2004.
[10] P. Stone, *Machine Learning*, Cambridge, MA: MIT Press, 2018.
[11] D. Jurafsky and J. H. Martin, *Speech and Language Processing*, Upper Saddle River, NJ: Pearson, 2019.
[12] H. Simon, *The Sciences of the Artificial*, Cambridge, MA: MIT Press, 1996.
[13] R. Murphy, *Introduction to AI Robotics*, Cambridge, MA: MIT Press, 2000.
[14] Y. Nakamura, "Flexible Manufacturing Systems: Recent Developments," *Annual Review of Control, Robotics, and Autonomous Systems*, vol. 3, pp. 1–23, 2020.
[15] M. Hatakeyama et al., "The ATRA Project: Advanced Robotics Technology and Their Applications," *IEEE Robotics & Automation Magazine*, vol. 25, no. 2, pp. 44–52, 2018.
[16] R. P. Bonasso, R. Murphy, and T. L. Johnson, Experiences with an Architecture for Intelligent, Reactive Agents. In *Proceedings of the Second International Conference on Principles of Knowledge Representation and Reasoning*, Cambridge, MA: KR'91, 1991, pp. 145–156.
[17] A. Samuel, "Some Studies in Machine Learning Using the Game of Checkers," *IBM Journal of Research and Development*, vol. 3, no. 3, pp. 210–229, 1959.
[18] T. Mitchell, *Machine Learning*, New York, NY: McGraw-Hill, 1997.
[19] I. Goodfellow, Y. Bengio, and A. Courville, *Deep Learning*. USA: MIT Press, 2016.
[20] L. Fei-Fei, R. Fergus, and P. Perona, "Learning Generative Visual Models from Few Training Examples: An Incremental Bayesian Approach Tested on 101 Object Categories," *Computer Vision and Image Understanding*, vol. 106, no. 1, pp. 59–70, 2007.
[21] A. Krizhevsky, I. Sutskever, and G. E. Hinton, Imagenet Classification with Deep Convolutional Neural Networks. In *Advances in Neural Information Processing Systems,* USA: MIT Press, 2012, pp. 1097–1105.
[22] R. S. Sutton and A. G. Barto, *Reinforcement Learning: An Introduction,* USA: MIT Press, 2018.
[23] R. Reiter, *Knowledge in Action: Logical Foundations for Specifying and Implementing Dynamical Systems*, USA: MIT Press, 2001.
[24] N. Chakraborty, S. Kar, and S. Ghosh, A Review on Industrial Automation Process. In *2016 International Conference on Electrical, Electronics, and Optimization Techniques (ICEEOT)*, Chennai, India, 2016, pp. 2940–2943.
[25] A. A. Souza and R. Borenstein, Mobile Robot Localization Using Landmarks. In *Proceedings of the IEEE International Conference on Robotics and Automation*, San Francisco, CA, USA, April 2000, pp. 2831–2836.

[26] M. Quigley et al., "ROS: An Open-Source Robot Operating System," *IEEE International Conference on Robotics and Automation (ICRA)*, 2009.
[27] S. Thrun et al., Robotic Mapping: A Survey. In *Exploration of Mobile Robots*, Singapore: Springer, 2002, pp. 1–29.
[28] A. Geiger, P. Lenz, and R. Urtasun, "Are We Ready for Autonomous Driving? The KITTI Vision Benchmark Suite," *IEEE Conference on Computer Vision and Pattern Recognition (CVPR)*, 2012.
[29] D. Silver et al., "Mastering the Game of Go with Deep Neural Networks and Tree Search," *Nature*, vol. 529, no. 7587, pp. 484–489, 2016.
[30] S. Thrun, W. Burgard, and D. Fox, *Probabilistic Robotics*, USA: MIT Press, 2005.
[31] R. Szeliski, *Computer Vision: Algorithms and Applications*, Singapore: Springer, 2010.
[32] Y. Wang et al., "Internet of Things (IoT) Enabled Real-Time Energy Management Framework for Smart Buildings," *IEEE Access*, vol. 8, pp. 40831–40843, 2020.
[33] D. Hanson et al., "Robotic Machines with Emotionally Expressive Faces," U.S. Patent No. 9,624,578, 2017.
[34] J. Lee et al., "Artificial Intelligence and Robotics and Their Impact on the Workplace," *SSRN Electronic Journal*, vol. 43, no. 2, pp. 178–191, 2018.
[35] Y. Zhang et al., "Integration of Artificial Intelligence and Robotics in Manufacturing: A Review," *IEEE Transactions on Industrial Informatics*, vol. 15, no. 6, pp. 3439–3450, 2019.
[36] M. Liu et al., "Enhancing Manufacturing Operations Using AI-Driven Robotics in Industry 4.0," *IEEE Transactions on Automation Science and Engineering*, vol. 17, no. 2, pp. 925–936, 2020.
[37] S. Wang et al., "Predictive Maintenance in Industry 4.0: A Review," *IEEE Access*, vol. 8, pp. 155910–155924, 2020.
[38] S. N. Yoon and D. Lee, "Artificial Intelligence and Robots in Healthcare: What Are the Success Factors for Technology-based Service Encounters?" *Technological Forecasting and Social Change*, vol. 152, p. 119814, Feb. 2020. Available: https://doi.org/10.1016/j.techfore.2019.119814
[39] W. Wallach and C. Allen, *Moral Machines: Teaching Robots Right from Wrong, Oxford University Press,* 2009. Available: https://global.oup.com/academic/product/moral-machines-9780195374049?cc=us&lang=en&.
[40] T. Kurzweil, "GPT-4 and beyond: The future of language understanding in AI," Singularity Hub, Dec. 2023. [Online]. Available: https://singularityhub.com/2023/12/15/gpt-4-and-beyond-the-future-of-language-understanding-in-ai/. [Accessed: Feb. 14, 2024].
[41] C. Eliasmith, *How to Build a Brain: A Neural Architecture for Biological Cognition,* Oxford University Press, 2013.

Chapter 3

Programming platforms of autonomous robotic systems

Ngum Breayon Kwe

3.1 INTRODUCTION TO AUTONOMOUS ROBOTIC SYSTEMS

Autonomous robotic systems represent a long-term evolution paradigm in the field of robotics, embodying the fusion of advanced hardware, sophisticated software, and artificial intelligence to enable machines to operate independently and adapt to dynamic environments [1, 2]. These systems, are often equipped with leveraging sensors, actuators, and intelligent algorithms, which have the capability to perceive their surroundings, make optimized decisions, and execute tasks without any or partial continuous human intervention [3–5]. The evolution of autonomous robotics has far-reached implications across diverse industries, from manufacturing and healthcare to exploration and transportation [6–8].

3.1.1 Fundamentals of robotic programming

Robot programming is the backbone of every autonomous robotic system, encompassing the set of instructions that govern a robot's actions [9, 10]. Understanding the fundamentals of robot programming is very important for developing autonomous systems, which can effectively execute tasks in dynamic environments.

Low-level languages: assembly language and C are often used for programming the microcontrollers and processors embedded in robotic systems. These languages provide direct control over hardware but can be complex to use.

High-level languages: Python, Java, and C++ are popular object-oriented programming high-level languages used in robotics. They offer abstraction and ease of development, facilitating rapid prototyping and algorithm implementation.

Procedural programming: in procedural programming, the focus is on procedures or routines that define a series of steps to be executed. It

DOI: 10.1201/9781003537083-3

is well-suited for tasks with clear sequential steps, such as robotic arm movements for assembly.

Reactive programming: it focuses on real-time responsiveness to environmental changes. Reactive systems continuously react according to sensor inputs without a predefined sequence of steps.

Understanding these programming paradigms will equip researchers and robotic programmers with the tools to design efficient and scalable software architectures for autonomous systems.

3.1.2 Integration of programming with hardware

Middleware: middleware are facilities that enable communication between the hardware and software components of a robotic system. The Robot Operating System (ROS) is a widely used middleware that provides a framework for developing robotic software.

Real-time operating systems (RTOS): in some applications, most often those requiring precise timing, real-time operating systems are deployed to ensure that tasks are executed within specified time constraints. There are a few simulation and testing systems used for RTOS.

Simulations: before robotic systems are deployed in real-world environments, developers often use simulations to test algorithms and their behaviors. Simulation tools like Gazebo and V-REP enable realistic virtual testing.

Hardware-in-the-loop (HIL) testing: this involves connecting the robot's hardware to a simulation, allowing developers to test the system's interaction between software and physical components.

3.2 HARDWARE PLATFORMS FOR AUTONOMOUS ROBOTS

Autonomous robots rely on a diverse array of hardware components to interact with their environment, process information, and execute tasks. The choice of hardware platforms to use is crucial, it influences a robot's capabilities, adaptability, and overall performance. Below is a detailed explanation of the main hardware components commonly found in autonomous robotic systems.

- **Microcontrollers and microprocessors**

Microcontrollers and microprocessors serve as the central processing units (CPUs) of robotic systems, responsible for executing algorithms and making decisions. Often embedded in smaller robotic systems, microcontrollers like Arduino, STM32, are designed for real-time control. They have built-in memory, input/output peripherals, and are well-suited for tasks that require simplicity and low power consumption. But on the other hand, platforms like Raspberry Pi or BeagleBone offer more computational complexity, power, support for operating systems, and broader connectivity options. These processors are suitable for applications demanding higher-level processing and complex algorithms.

- Sensors

Sensors provide robots with the ability to perceive their surroundings by capturing data from the environment. A variety of sensors are deployed in autonomous robots, most often used sensors include the following: camera sensors, LiDAR (Light Detection and Ranging), inertial measurement units (IMUs), etc.

Camera sensors: cameras provide vision capabilities, enabling robots to interpret and understand their surroundings. Advanced vision systems, like stereoscopic cameras or depth sensors, enhance perception accuracy.

LiDAR: sensors that emit laser beams to measure distances accurately. They are crucial for mapping environments in three-dimensional coordinates and ensuring obstacle avoidance.

IMUs: inertial measurement units, combining accelerometers and gyroscopes, providing information about a robot's orientation, acceleration, and angular velocity.

- Actuators

Actuators are mechanisms responsible for converting electrical signals into physical movements, which could be linear or rotational motion. Examples of these actuators include motors and servos.

3.3 SIMULATIONS AND VIRTUAL ENVIRONMENTS

Simulations and virtual environments are computer-generated representations that mimic real-world scenarios, providing a virtual platform for testing and validating robotic algorithms and behaviors before deploying them on physical robots.

3.3.1 Key importance of simulations and virtual environments

- **Risk mitigation:** simulations permit developers to assess the performance of algorithms in various scenarios without the risk of damaging physical robots or causing harm to itself.
- **Algorithm development:** developers can iterate rapidly in a virtually simulated environment, fine-tuning algorithms and strategies before implementation on real hardware.

- **Cost savings:** simulations reduce the costs associated with physical prototyping and testing, making them an economical solution for algorithm validation.

3.3.2 Simulation platforms

A. Gazebo: gazebo is an open-source robotics simulator that provides a realistic and physics-based 3D environment. It supports the simulation of robots, sensors, and even entire robot environments.

3.3.2.1 Key features of Gazebo

- Gazebo's physics engine simulates interactions between robots and their surroundings accurately.
- It supports the simulation of various sensors, including cameras, LiDAR, and IMUs.
- Gazebo seamlessly integrates with the Robot Operating System (ROS), allowing programmers and developers to test ROS-based algorithms in a simulated environment.

B. V-REP (Virtual Robot Experimentation Platform): V-REP is a versatile robot simulation platform with a user-friendly interface. It allows users to create and simulate robotic scenarios using a wide range of predefined robots and sensors.

3.3.2.2 Key features of V-REP

- V-REP offers a comprehensive library of robotic models, making it suitable for various applications.
- The remote API enables communication between V-REP and external applications, facilitating integration with control algorithms.
- V-REP supports the simulation of multiple robots interacting in the same environment.

C. Webots: this is a professional robot simulation software used for modeling, simulation, and control of robots. It provides a 3D physics engine and a user-friendly interface.

3.3.2.3 Key features of Webots

- Webots incorporate a robust physics engine for realistic interactions between robots and their environment.
- It is compatible with various operating systems, including Windows, Linux, and macOS.

- Webots allows users to generate code for robot controllers, making it easier to transition from simulation to real-world implementation.

3.3.3 Applications simulation and virtual environments for autonomous robots

- Simulations enable developers to test navigation algorithms, object detection, and path planning in a controlled and repeatable environment.
- Simulated environments are invaluable for testing algorithms that involve multiple robots collaborating or interacting with each other.
- Simulations can be integrated into HIL setups, allowing developers to test algorithms with real hardware components in a simulated environment.

3.3.4 Limitations and considerations

- Striking a balance between realistic simulations and computational efficiency is a challenge, as more realism often requires more computational resources.
- Simulating sensor data, especially for complex sensors like LiDAR and cameras, may not perfectly replicate real-world conditions.
- While simulations provide a valuable testing ground, the transition from simulation to the real world can still present challenges due to differences in dynamics and environmental factors.

3.3.5 Future scope and future trends

- Simulations are increasingly being used for training machine learning models for robotic applications, leveraging the advantages of synthetic data.
- Cloud-based simulation platforms allow distributed and collaborative testing, making it easier for teams to work on simulations from different locations.
- Integration with augmented reality (AR) and virtual reality (VR) technologies is becoming more prevalent, enhancing the user experience and making simulations more immersive.

Simulations using virtual environments play a crucial role in the testing and development phase of robotic systems, offering a safe, efficient and cost-effective means to refine algorithms and assess their performance before deploying them in real-world scenarios.

3.4 OPERATING SYSTEMS FOR ROBOTICS

A. Real-Time Operating Systems (RTOS)

RTOS is an operating system specifically designed to meet the requirements of real-time systems, ensuring that tasks are executed within specified time constraints. Real-time operating systems (RTOS) are used in environments where a dense number of events, mostly external to the computer operating system, must be accepted and processed in a short time or within certain deadlines. Such applications are industrial control, telephone switching equipment, flight control, and real-time simulations. With an RTOS, the processing time is measured in tenths of seconds (1/10)s. This system is time-bound and has a fixed deadline. The processing in this type of system must occur within the specified timely constraints. Otherwise, this will lead to system failure.

Examples of real-time operating systems are airline traffic control systems, airline reservation systems, heart pacemakers, command control systems, network multimedia systems, robotic systems, etc.

Figure 3.1 illustrates real-time operating systems can be of three types –

- **Hard real-time operating system:** these operating systems provide the ability to perform critical tasks are completed within a range of

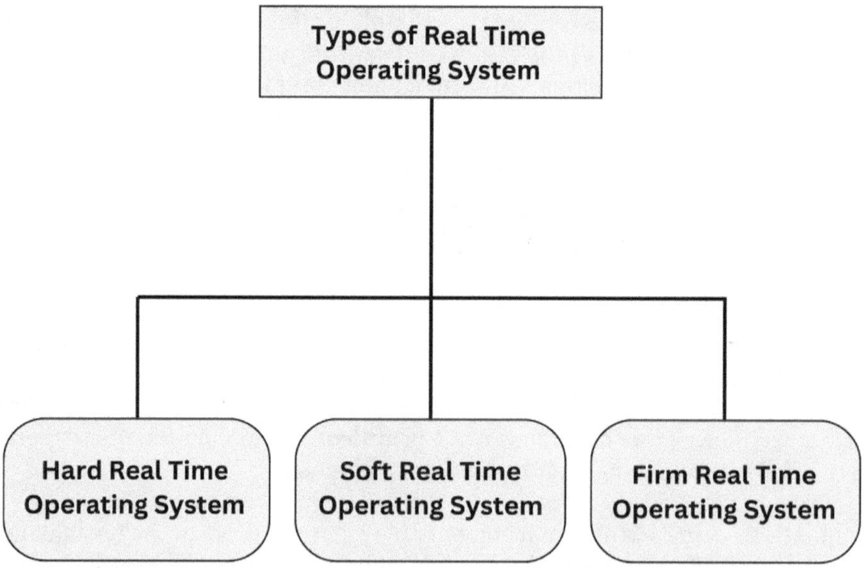

Figure 3.1 Architecture of the real-time operating system. (www.geeksforgeeks.org/real-time-operating-system-rtos/)

time. Imagine a robot is hired to weld a car body. If the robot welds too early or too late, the car cannot be sold, so it is a hard real-time system that requires complete car welding by the robot hardly on time, medical imaging systems, industrial control systems, weapon systems, scientific experiments, robot systems, air traffic control systems, etc.

- **Soft real-time operating system:** this SRTOS provides an extended bandwidth in the time limit in HRTOS. Practical examples include multimedia systems, digital audio systems, etc. Programmer-defined, explicit, and controlled processes are encountered in real-time systems. A separate process is changed by handling a single external event. The process is activated upon the occurrence of the related event signaled by an external interrupt. Multitasking operation is accomplished by scheduling processes for execution independently of one another. Every process is assigned a certain level of priority that corresponds to the relative importance of the event that it services. The required processor is allocated to the highest-priority processes in the system. This type of schedule is called priority-based preemptive scheduling.
- **Firm real-time operating system:** this RTOS type have to follow time constraints as well. In spite of its little impact, missing a deadline can have unintended consequences, which includes a reduction in the quality of the product. For example, multimedia applications.
- **Deterministic real-time operating system:** consistency is the main key in this type of real-time operating system. It ensures that every task and all processes are executed with predictable timing all the time. Due to this, this type of RTOS it more suitable for applications in which timing accuracy is very important. For example, INTEGRITY, PikeOS.

3.4.1 Applications of RTOS

- **Robotics control:** RTOS is essential for tasks that demand precise timing, such as sensor data processing, motor control, and feedback loops.
- **Safety-critical systems:** in applications where safety is of high priority, such as medical robots or autonomous vehicles, RTOS helps maintain control and responsiveness.

B. Linux-Based Systems

The Linux Operating System is a type of operating system that is similar to the Unix operating system, and it is built upon the Linux Kernel illustrated in Figure 3.2. The Linux Kernel is like the brain of the operating system because

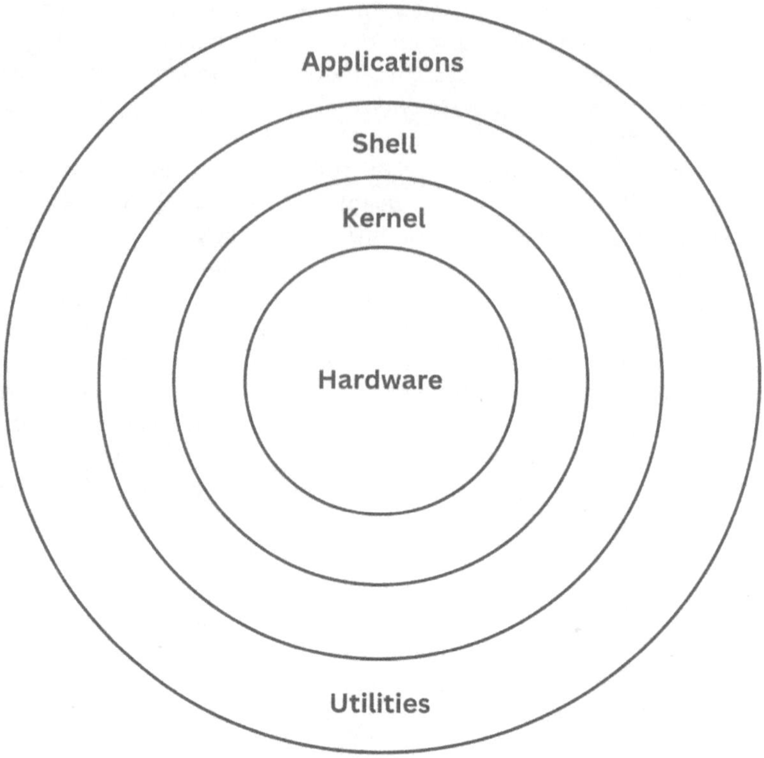

Figure 3.2 Linux operating system architecture. (www.geeksforgeeks.org/introduction-to-linux-operating-system/)

it manages how the computer interacts with its hardware and resources. It makes sure everything works smoothly and efficiently. But the Linux Kernel alone is not enough to make a complete operating system. To create a full and functional system, the Linux Kernel is combined with a set of software packages and utilities, which are together called Linux distributions. These distributions make the Linux Operating System ready for users to run their applications and perform tasks on their computers systems securely and effectively. Linux distributions come in different dynamics, each tailored to meet the specific needs and preferences of users.
Linux architecture has the following components:

- **Kernel:** the kernel is the core of the Linux-based operating system. It virtualizes the common hardware resources of the computer to provide each process with its virtual resources. This makes the process seem like it is the sole process running on the machine. The kernel is responsible for preventing and mitigating conflicts between different software processes. Different types of the kernels are monolithic kernel, hybrid kernels, exo kernels, micro kernels.

- **System library:** linux makes use of system libraries also known as shared libraries, to execute various functionalities of the operating system. These libraries contain predefined code that applications can actualize to perform specific tasks. By using these libraries, developers can save time and effort, as there is no need to rewrite the same code repeatedly. System libraries act as an interface between applications and the kernel, providing a standardized and efficient way for applications to interact with the underlying system.
- **Shell:** the shell is the user interface of the Linux Operating System. It enables users to interact with the system by entering commands, which is interpreted and executed by the shell. The shell serves as a bridge between the user and the kernel, forwarding the user's requests to the kernel for processing. It provides a convenient and effective way for users to perform various tasks, such as managing files, running programs, and configuring the system.
- **Hardware layer:** the hardware layer is the support layer for all the physical components of the computer, such as the HDD (Hard Disk Drive), CPU (Central Processing Unit), RAM (Random Access Memory), and input/output devices. This layer is responsible for interacting with the Linux Operating System and providing the necessary resources for the system and applications to function properly. The Linux kernel and system libraries enable communication and control over these hardware components, thus ensuring that they work harmoniously together.
- **System utility:** these are essential tools and programs provided by the Linux Operating System to manage and configure different aspects of the system. These utilities perform tasks such as installing software, monitoring system performance, configuring network settings, managing users and permissions, etc. System utilities simplify system administration tasks, making it easier for users to maintain their Linux systems efficiently.

3.4.2 Applications of Linux-based systems

- **General-purpose robotics:** Linux is suitable for a broader range of robotic applications, providing a flexible and customizable platform for development.
- **Research and development (R&D):** many research-oriented robotic projects, leverage Linux for its extensive support and community-driven development.

C. Integration with Robot Operating System (ROS)

ROS is not an operating system in the traditional sense but rather a middleware framework designed to simplify and standardize robotic software development. Robot Operating System or simply ROS is a framework which is used by hundreds to thousands of companies and techies of various fields all across the globe in the field of robotics and automation. It provides a painless and effortless entry point for nonprofessionals in the field of programming robots.

3.4.3 Applications ROS

- **Modular robotics:** ROS enhances the development of modular and reusable robotic software components, promoting collaboration and code sharing.
- **Communication and coordination:** ROS provides communication tools that allow different modules (nodes) to exchange information, making it easier to coordinate complex robotic tasks.

D. Real-Time Linux

Real-time Linux is a specialized operating system kernel designed to provide deterministic and predictable behavior for time-sensitive applications. Unlike traditional Linux distributions, where processes and tasks are scheduled with varying degrees of priority and time-sharing, real-time Linux ensures that critical tasks receive immediate attention and execute within predefined time constraints.

3.4.4 Applications of real-time Linux

- **Hybrid systems:** real-time Linux can be beneficial in applications where real-time capabilities are needed alongside the advantages of a full-fledged Linux environment.
- **Industrial automation:** real-time Linux is used in robotics for precise control and coordination in industrial automation processes.

E. Windows-Based Systems

Windows operating systems, such as Windows 10 IoT, are sometimes employed in robotic applications, especially in environments where Windows development tools are prevalent. Windows may be a commissioned OS within which ASCII text file is inaccessible. It is designed for people with no programming information, for business, and alternatively industrial users. It is very straightforward and simple to use. The distinction between the Linux and Windows package is that Linux is completely freed from price whereas Windows is a marketable package and is expensive.

3.4.5 Applications of Windows-based systems

- **Human–robot interaction:** Windows-based systems are used in applications where seamless interaction with human users, such as in service robots, educational robots, etc.
- **Custom applications:** some robotic systems with custom interfaces or applications built on Windows platforms leverage existing development tools.

F. Embedded Operating Systems

Every embedded system is task specific. They mostly do a particular task in loop/repeatedly for their entire lifespan. These systems are designed to execute their task within a particular time interval, and thus they have to be fast enough to be up to their time limit. Figure 3.3 shows the embedded operating system architecture. They have little or no user interface like a fully automatic washing machine does its task fully once its program is set and stops after its work is finished with almost no user interface. They are built to achieve a particularly good efficiency level. They have a very small operating system and need little power.

These systems cannot at all be upgraded or updated. Thus, they must be really high on efficiency and reliability since they cannot be updated.

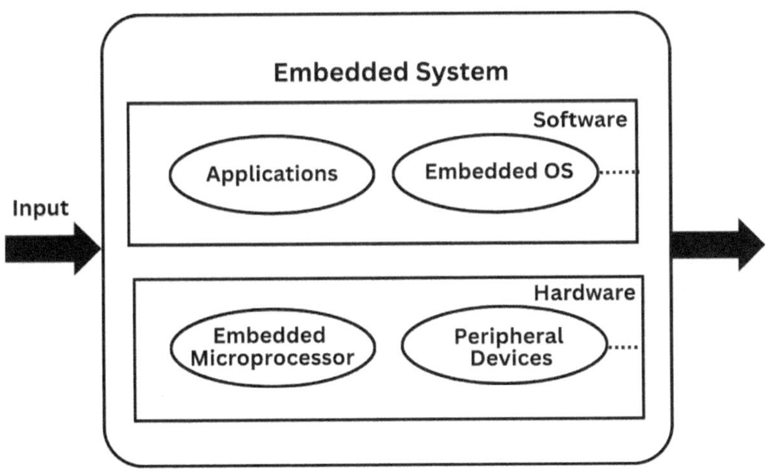

Figure 3.3 Architecture of the embedded operating system. (www.tutorialspoint.com/what-is-an-embedded-operating-system/)

3.4.6 Applications of embedded operating systems

- **Microcontroller-based systems:** in limited and resource-constrained robotic platforms, lightweight embedded operating systems provide the necessary support for basic control and communication.
- **IoT devices:** embedded operating systems are suitable for robotic applications that are part of the Internet of Things (IoT), where low power consumption and efficient resource utilization are crucial.

G. Safety-Critical Operating Systems

A safety critical RTOS is a real-time operating system that was designed to integrate with a specific safety critical design standard. Compliance to these standards guarantees performance and creates a robust and deterministic RTOS, allowing the safety critical RTOS to be used widely in industries where failure could result in harm to people – e.g., medical (IEC 62304), industrial applications (IEC 61508) automotive (ISO 26262), and aerospace (DO-178C).

3.4.7 Applications of a safety-critical operating system (SCOS)

- **Medical robotics:** in surgical robots and other medical applications, safety-critical operating systems play a vital role in ensuring the reliability of robotic systems during critical procedures.
- **Aerospace and defense:** autonomous systems in aerospace and defense often require safety-critical operating systems to guarantee reliable operation in challenging and dynamic environments.

Choosing the right operating system for a robotic system depends on factors such as the specific requirements of the application, the level of real-time performance needed, the availability of development tools, and the preferences of the development team. The diverse landscape of operating systems for robotics reflects the wide range of applications and use cases within the field.

3.5 SENSOR FUSION AND PERCEPTION: IN-DEPTH ANALYSIS

A. Sensor Fusion

Sensor fusion is a technique used to combine data from multiple sensors to provide a more complete and accurate representation of its environment or the system being monitored. The aim or idea here is to use the strengths of each sensor to compensate for the weaknesses of others, resulting in a more robust and reliable system. This is crucial in robotics to enhance perception and decision-making.

3.5.1 Types of sensor fusion

There are several types of sensor fusion, including the following:

1. **Data fusion:** this type of sensor fusion combines data from multiple sensors at the raw data level. The goal is to improve the accuracy and precision of the data by combining information from multiple sources.
2. **Feature fusion:** this type of sensor fusion combines multiple sensors data at the feature level. The goal is to extract relevant features from each sensor and combine them to create a more comprehensive/understandable representation of the environment.
3. **Decision fusion:** this type of sensor fusion combines the decisions/conclusions made by multiple sensors. The goal is to improve the overall decision-making process by taking into account multiple sources of information.
4. **Multi-modal sensor fusion:** this type of sensor fusion combines data from multiple sensors that measure different physical phenomena. For example, it can use data from a camera and a LiDAR sensor to create a more complete representation of the environment.
5. **Hierarchical sensor fusion:** this type of sensor fusion is a process of combining information from multiple sensors at different levels in the system, such as at sensor level, feature level, and decision level.

These are just a few examples of the various existing types of sensor fusion, but the possibilities are almost endless as technology makes more advancement in future. Sensor fusion can be applied to a broader range of applications including robotics, autonomous systems, medical devices, etc.

B. Perception Algorithms

These algorithms play a crucial role in helping robots comprehend their surroundings. They enable robots to collect and interpret data from various sensors like cameras, RGBD, and LiDAR sensors. Through this data processing, robots can identify objects, track objects, measure distances, and navigate around obstacles. In essence, perception algorithms provide robots with the ability to gain control of and understand their environment. Furthermore, enabling them to interact with it safely and effectively. These algorithms are used to process raw sensor data to extract meaningful features, which facilitates task execution like object recognition, mapping, and navigation.

3.5.2 Classification of perception algorithms

1. Perception of Static Obstacles
In the domain of perceiving static obstacles, robots employ sensors like LiDAR or RGBD to detect objects within their environment, but these algorithms typically provide the robot with limited information. While they can ascertain the relative position of static obstacles, they lack semantic information about these objects. This implies that the robot knows where an obstacle is but lacks details like the object's identity (e.g., chair, table) and any prediction of its future movements in its environment. Consequently, when dealing with dynamic obstacles such as moving pedestrians, robots relying solely on static perception will treat them as immobile objects. This can lead to potential collisions, as the robot does not take account of the dynamic nature of these obstacles.

2. Perception of Dynamic Obstacles
Dynamic obstacles introduce an additional layer of complexity to the robot's perception task. Using sensors like RGBD, LiDAR, and cameras. In this case, robots not only detect these obstacles but also gain valuable semantic insights about the obstacle's category (e.g., chair, pedestrian, table). Armed with this information, robots can distinguish between static and dynamic obstacles. For dynamic obstacles, robots employ continuous multi-frame tracking to estimate their speed and direction. Thus by predicting the future path of these dynamic obstacles, robots can proactively plan their paths to avoid collisions.

3.5.3 Challenges in sensor fusion and perception

- Sensors are prone to noise and uncertainties. Sensor fusion algorithms must account for these factors to avoid inaccuracies in the fused data.
- Achieving real-time processing is a huge challenge, especially when dealing with large datasets from multiple sensors. Efficient algorithms and hardware acceleration are employed to tackle this challenge.
- Changes in lighting conditions, weather, or the presence of dynamic obstacles pose challenges to perception systems. Adaptive algorithms are designed to handle environmental variability.

3.5.4 Applications in robotics

- **Autonomous navigation:** sensor fusion enhances the accuracy of localization and mapping, enabling robots to navigate autonomously in dynamic environments.

- **Object recognition and manipulation:** perception algorithms contribute to object recognition, allowing robots to identify and manipulate objects with precision.
- **Human—robot interaction:** advanced perception capabilities enhance human–robot interaction by enabling robots to recognize and respond to human gestures, expressions, and commands.

Moreover the challenges, the advance applications of sensor fusion and perception in robotics continually evolves. Advances in algorithms, machine learning integration, and the integration of diverse sensor technologies contribute to the ongoing enhancement of robotic perception capabilities. The ability to robustly perceive and interpret its environment is fundamental and crucial for the successful deployment of autonomous robotic systems across various applications.

3.6 LOCALIZATION AND MAPPING: IN-DEPTH ANALYSIS

A. Localization

Robot localization is the process of determining where a mobile robot is positioned with respect to its environment. It involves estimating the robot's pose (position and orientation), typically represented as its position (x, y) and orientation (theta or yaw angle). The knowledge of the robot's own location is an essential precursor to making decisions about future actions. Robotic mapping is a discipline outsourced from computer vision and cartography. Through the integral of localization and mapping, the goal for an autonomous robot is to construct a map or floor plan in order to locate itself in its environment.

3.6.1 Localization techniques

- **Odometry:** this is making use of motion sensors to determine the robot's change in position relative to some known position. The fundamental concept behind this technique is the incremental change in position over time. The change in position called linear displacement relative to the floor, can be measured on the basis of revolutions of the wheel. This method provides good accuracy for the short-term measurement, but this technique is prone to error when done on a significant displacement.
- **Sensor fusion:** sensor fusion is the process of combining sensor data or data derived from dynamic sources such that the resulting information has less uncertainty than would be possible when these sources are used individually.

- **Visual localization:** visual localization, a technique to determine location of the robot via images. This method is most applicable in self-driving cars to estimate their position, used in augmented reality applications to interact with the physical world both indoor and outdoor.

B. Mapping

Mapping involves creating a representation of the environment that the robot can use for navigation that is a 3D configuration space. The map can be 2D or 3D and may include information about obstacles, landmarks, or the terrain. Mapping creates a spatial model of the environment surrounding the robot using its sensors. The map is then used for localization and navigation. Figure 3.4 illustrates a block diagram of the different mapping techniques used.

3.6.2 Mapping techniques

- Sensorial Maps

Robots make use of sensors to perceive their environment. This map is a direct representation of the robots environment based on quality and accuracy. Robotics combines the measurements obtained via odometry information

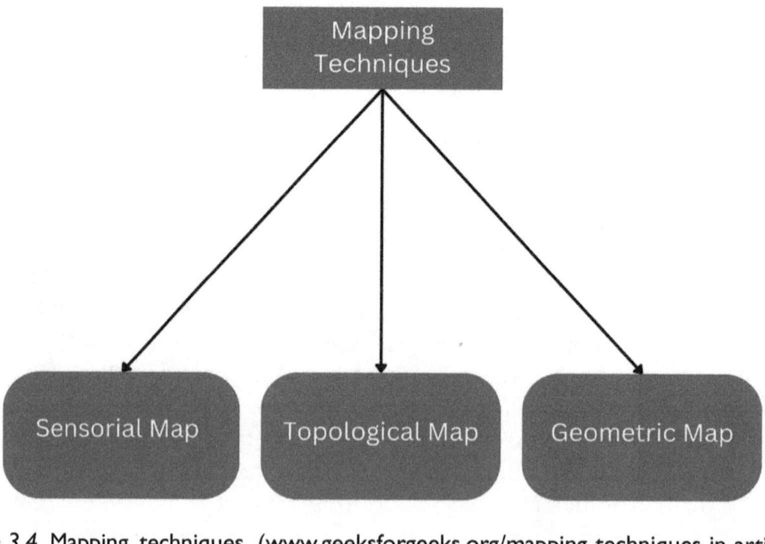

Figure 3.4 Mapping techniques. (www.geeksforgeeks.org/mapping-techniques-in-artificial-intelligence-and-robotics/)

and integrate techniques like servo control to identify the sensor responses to navigate the corresponding map. When the robot moves throughout the environment it collects sensor measurements, which we interpret to form a map for the robot's environment for accurate navigation.

- Topological Maps

The long metric maps are difficult to maintain, an alternative to it would be topological maps that are a graph-based representation. Wherein the nodes represent the places and the edges represent the arcs/paths as seen in Figure 3.5.

- Geometric Maps

Geometric maps are well-organized information of the robot's environment as they are represented by a coordinates' system. The inspection of unidentified environment results in geometric maps from discovery and related problems, which are studied largely in the area of computational geometry. These maps are sensitive to noise but every map object is linked radially to develop an image of the open space. The radial polygons are used for localization.

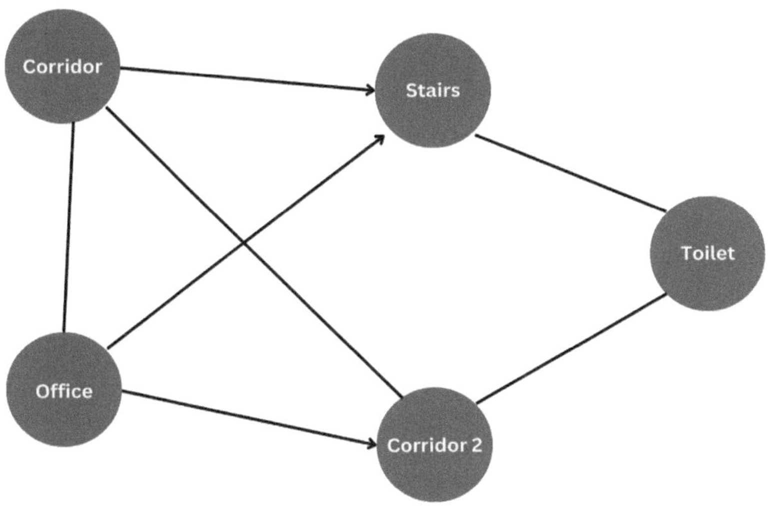

Figure 3.5 Topological maps. (www.geeksforgeeks.org/mapping-techniques-in-artificial-intelligence-and-robotics/)

C. Simultaneous Localization and Mapping (SLAM)

SLAM is an estimation of both the pose of a robot and the map of the environment simultaneously. SLAM is hard to implement because a map is needed for localization and a good pose estimate is needed for mapping. Key Components

- **Localization:** inferring location given a map.
- **Mapping:** inferring a map given locations.
- **SLAM:** learning a map and locating the robot simultaneously.

Localization and mapping are foundational aspects of robotic systems, which essentially enables robots to understand their position and surroundings. Advances in algorithms, sensor technologies, and SLAM techniques contribute to more robust and accurate localization and mapping capabilities, opening up possibilities for a broad range of applications in robotics.

3.7 PLANNING AND DECISION-MAKING

Planning and decision-making in robotics refer to the process by which a robot determines a sequence of actions to achieve a specific goal or navigate through its environment. Motion planning and decision-making are at the core of robotics and AI. In theory, these problems can be solved using optimal control or dynamic programming. However, computational cost for solving many real world problems is prohibitively high and is exacerbated by the "curse of dimensionality." Randomized sampling-based methods (like RRT*) ameliorate this problem by avoiding priori griding of the environment and incrementally building a graph online. It involves both high-level planning, defining the overall strategy, and low-level decision-making, determining the specific actions to execute.

3.7.1 Key steps involved in high-level planning

- **Strategic decision-making:** at a higher level, robots need to make strategic decisions about their overall goals and objectives. This could include tasks like determining the best path to a destination, identifying targets, or selecting optimal actions for a given scenario.
- **Path planning:** in navigating tasks, robots use path planning algorithms to find the most optimal route from their current position to a target location. These algorithms consider factors such as obstacles, terrain, and possibly dynamic elements in the environment.
- **Task planning:** for more complex tasks, robots engage in task planning, breaking down a larger goal into a sequence of sub-tasks.

This involves decision-making about the order and execution of actions to accomplish the overarching objective.

3.7.2 Low-level decision-making

- **Local environment interaction:** at a lower level, robots continuously make decisions based on real-time interactions with their immediate environment. This involves obstacle avoidance, adapting to dynamic changes, responding to sensor inputs, etc.
- **Reactive control:** reactive control strategies permit robots to respond rapidly to changes in the environment. These systems make decisions based on immediate sensor feedback, allowing for rapid adjustments in movement or behavior.
- **Feedback loops:** decision-making is often embedded in feedback loops where the robot constantly assesses its current and previous actions, comparing them to the desired outcomes, and adjusts its behavior accordingly.

Planning and decision-making are fundamental aspects of robotic systems, enabling them to navigate, interact, and accomplish tasks in dynamic environments.

3.8 COMMUNICATION PROTOCOLS FOR ROBOTICS

Communication protocols have always been a big challenge for the robotic industry. Protocols are used to exchange and share data between two devices. In industrial applications, they are often used to enable a computer to communicate with a robot or an end effector. In order to be able to communicate, the two devices need to understand the selected protocol. These protocols enable seamless interaction and coordination among various subsystems, actuators, sensors, controllers, etc.

3.8.1 Types of communication protocols

There are two types of communication protocols which are classified as follows:

1. **Inter-System Protocol**

The inter-system protocol is used to communicate between two different devices as illustrated in Figure 3.6. Like communication between computer to microcontroller kit and PC. The communication is done through an inter-bus system.

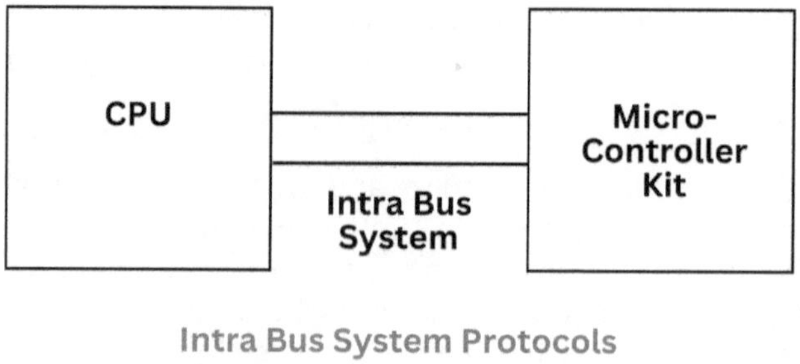

Figure 3.6 Inter-bus system protocol. (www.elprocus.com/communication-protocols/)

The different categories of the inter-system protocol mainly include the following:

- UART Protocol

UART stands for a universal asynchronous transmitter and receiver. A UART protocol is a serial communication with the commonly used two wired protocols. The data cable signal lines are labeled as Rx and Tx. Serial communication is commonly used for transmitting and receiving the signal between the connected devices. It is transferred and receives the data serially bit by bit without class pulses. The UART takes bytes of data and sends them in individual bits in a sequential manner.

The UART is a half-duplex protocol. Half-duplex implies transferring and receiving data but not at the same time. Most of the controllers have hardware UART on board. It uses a single data line for transmitting and receiving the data. It has one start bit, 8-bit data and a one-stop bit. This means the 8-bit data transfers one's signal is high to low. Examples, emails, SMS, walkie-talkies, etc.

- USART Protocol

USART stands for a universal synchronous and asynchronous transmitter and receiver. It is a serial communication of the commonly used two-wire protocol. The data cable signal lines are labeled as Rx and TX. This protocol is used to transmitting and receiving the data byte by byte along with the synchronized clock pulses. It is a full-duplex protocol, which means transmitting and receiving data simultaneously to different baud rates. Different devices communicate with a microcontroller through this protocol. Example, -telecommunications, etc.

- USB Protocol

USB stands for universal serial bus. Again it is a serial communication of two-wire protocol. The data cable signal lines are labeled D+ and D–. This protocol is applicable for communicating with the system peripherals. USB protocol is used to send and receive data serially to and from the host and peripheral devices. The USB communication requires driver software, which are based on the functionality of the system and they can transfer data on the bus without any request on the host computer.

Nowadays, most devices use the USB protocol communication technique. For example, like a computer to communicate with an ARM controller using USB. USB transfers data through different modes, the first one is the slow-speed mode 10 kbps to 100 kbps, the second is the full-speed mode 500 kbps to 10 mbps and the high-speed mode 25 mbps to 400 Mbps. The USB maximum cable length of 4 meters. Examples, mouse, hubs, switches, keyboard, pen drive, etc.

2. Intra-System Protocol

The intra-system protocol is used to communicate between two devices within the circuit board. While using these intra-system protocols shown in Figure 3.7, without going to inter-system protocols we will expand the peripherals of the microcontroller. The circuit complexity and power consumption will be enhanced by using the intra-system protocol. Using intra-system protocols' circuit complexity and power consumption, the cost is decreased, making this protocol quite efficient and secure for accessing data.

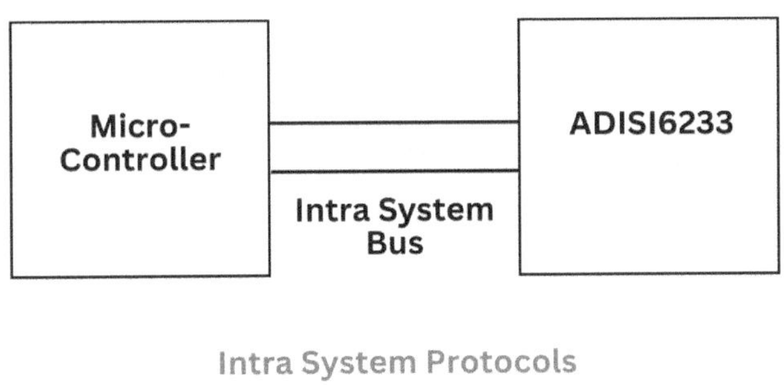

Figure 3.7 Intra-bus system protocol. (www.elprocus.com/communication-protocols/)

3.8.2 The different categories of intra-system protocol mainly include the following

- I2C Protocol

I2C stands for the inter-integrated circuit and it requires only two wires connecting all peripherals to the microcontroller. I2C makes use of two wires SDA (serial data line) and SCL (serial clock line) to transmit and receive information between devices. It is a master to a slave communication protocol. Each slave peripheral has a unique address. The master device sends the address of the target slave device and reads/writes the flag. If the address matches any slave device that the device is ON, the remaining slave devices are in disabled mode.

Once the address of any slave is matched, communication proceeds between the master and that slave device is enabled either for transmitting and receiving data. The transmitter sends 8-bit data, the receiver replies with 1-bit of acknowledgment. When the communication is completed, the master issues the stop condition. The I2C bus was developed by Philips Semiconductors. Its aim is to provide an easy way to connect CPU to peripherals chips.

- SPI Protocol

SPI stands for the serial peripheral interface. It is one of the serial communication protocol developed by Motorola. More often, the SPI protocol is also called a four-wire protocol. It requires four wires, which are MOSI, MISO, SS, and SCLK. The SPI protocol is used to communicate between the master and slave devices. The master first configures the clock using a frequency, it then selects the particular slave device for communication by pulling the chip select button. That particular device is then selected and starts the communication between the master and that particular slave. The master can only selects one slave at a time. It is a full-duplex communication protocol and it is not limited to 8-bit words in the case of bit transferring.

- CAN Protocol

CAN stands for the controller area network. It is a serial communication protocol, which requires two wires CAN High (H+) and CAN low (H-).

Communication protocols are the backbone of effective interaction within robotic systems. Whether wired or wireless, these protocols play a crucial role in enabling reliable, and real-time communication between the various components of a robotic system thus contributing enormously to the seamless operation in diverse robotic applications.

3.9 INTEGRATION OF AI IN ROBOTICS

AI-powered robots, mimicking human actions and intelligence, are revolutionizing the tech industries. This fusion of robotics and artificial intelligence

(AI) is reshaping manufacturing, healthcare, and more. As these mechanical entities become smarter, the synergy between integral AI and robotics continues to propel innovation and automation, making our world more efficient, and interconnected.

3.9.1 Key components responsible of AI integration in robotics

AI plays a crucial role in modern robotics, bringing intelligence and adaptability to these fascinating machine systems. Artificial intelligence (AI) is integrated in robotics in order to enhance the capabilities of robots, additionally enabling them to execute a wide range of tasks with increased autonomy and adaptability. There are several ways in which AI is applied in robotics:

1. Computer Vision
    - **Object recognition:** AI-powered computer vision allows robots to recognize and identify objects in their environment. Computer vision enhances the robots capability to understand their surroundings, create maps, and navigate through complex environments. This is essential for autonomous vehicles, drones, and robots operating in unstructured and dynamic spaces.
    - **Visual serving:** AI allows robots to track and precisely manipulate objects based on visual feedback, crucial for tasks like painting, welding, or assembling delicate components.

AI algorithms process camera and sensor data to map its surroundings, identify obstacles, and plan safe and efficient paths for robots to navigate.

2. Natural Language Processing (NLP)
    - **Human—robot interaction:** robots can understand and respond to natural language commands, enabling more intuitive and collaborative interactions with humans.
    - **Voice control:** robots can be controlled through voice commands, making them accessible for a broader range of users.
    - **Sentiment analysis:** AI can analyze human text and speech to understand their emotions and adjust robot behavior accordingly.
3. Machine Learning
    - **Autonomous decision-making:** AI algorithms can learn from driven data and make decisions in real-time, enabling robots to adapt to changing terrains and react to unexpected situations.
    - **Reinforcement learning:** robots can learn even machine and motor skills and control strategies through trial and error, enabling them to execute complex tasks like running, walking, playing games etc.

- **Predictive maintenance:** AI can analyze sensor data to predict equipment downtime and schedule preventive maintenance, reducing downtime and maintenance costs.

3.9.2 Applications of AI-integrated robotics

A. **In autonomous vehicles:** AI algorithms are essential for autonomous vehicles, enabling them to perceive the environment like roads, make decisions about navigation, and respond to dynamic scenarios. This is evident in self-driving cars, drones, and other autonomous systems.

B. **In robotic vision:** AI-powered vision systems enhance the visual perception of robots. Tracking, object recognition, and scene understanding are significantly improved, making robots more adept at executing precision tasks such as pick-and-place, quality inspection, and navigation.

C. **In adaptive manufacturing:** during manufacturing, AI-integrated robots can adapt to a large variation in production lines, handle dynamic tasks, and optimize processes. This adaptability is particularly valuable to industries with frequent product changes or customization requirements.

D. **Healthcare robotics:** AI contributes to robotic applications in healthcare in diverse sectors such as diagnostics, robotic surgery, and patient care. AI algorithms can assist in surgery, analyze medical data, and enhance the overall efficiency of healthcare robots.

E. **Service robots:** robots in service-oriented industries, such as hospitality and customer service, benefit from AI capabilities, which enable robots to understand natural language, navigate dynamic environments, and assist customers.

3.9.3 Challenges and considerations for AI integration

A. **Data quality and bias:** AI models heavily depend on the quality and diversity of training driven data. Ensuring unbiased and representative datasets is crucial to avoid perpetuating interference and bias in robotic decision-making.

B. **Explainability:** as AI algorithms become more dynamic and complex, there is an increasing need for explainability. Understanding how AI systems arrive at decisions is crucial, especially in critical applications where human safety is a concern.

C. **Security concerns:** the integration of AI introduces new cybersecurity challenges. Ensuring the security of AI algorithms is 100% and preventing malicious attacks on robotic systems is of greatest priority.

D. **Human–Robot collaboration:** harmonizing the collaboration between AI-driven robots and human workers requires attention to transparency, safety, and establishment of clear guidelines for shared tasks.

In the intricate integration of AI and robotics, our world is witnessing transformation as the field of robotics advances. From manufacturing to healthcare, the marriage of artificial intelligence and robotic systems is reshaping industries, ushering in a new era of unprecedented efficiency, adaptability, and autonomous capabilities. The synergy between these fields continues to redefine the possibilities and elevate technological landscapes.

3.10 CASE STUDIES AND PRACTICAL APPLICATIONS OF AI-INTEGRATED ROBOTICS

1. Autonomous Vehicles
   Case study: self-driving cars
   Application: companies like Tesla, Waymo, and Uber have been deploying AI-integrated robotic systems in self-driving cars for the past decade now. These vehicles use a combination of sensors and AI-generated algorithms for real-time perception, decision-making, and autonomous navigation.
   Practical impact: it has enhanced road safety, reduced rate of accidents, and enhanced the potential for more efficient transportation with reduced reliance on human drivers.
2. Manufacturing and Industry 4.0
   Case study: collaborative robots in manufacturing
   Application: companies like Universal Robots and ABB have long implemented collaborative robots (Cobots) into their manufacturing processes. These Cobots, equipped with AI, can adapt to different tasks, work alongside human operators and optimizing production efficiency.
   Practical impact: it increased flexibility in production lines, enhanced faster adaptation to changing demands, and improved human–robot collaboration in manufacturing arenas.
3. Healthcare Robotics
   Case study: surgical robots
   Application: systems like the da Vinci Surgical System leveraging AI for robotic-assisted surgery. The robots are controlled by a surgeon with the help AI algorithms to enhance optimal precision, minimize invasiveness, and improve surgical outcomes.
   Practical impact: more precise surgeries performed, reduced recovery time/delay, improved the ability to perform complex procedures through assistance of AI-driven robotic systems.
4. Logistics and Warehousing
   Case study: autonomous mobile robots (AMRs) in warehouses
   Application: companies like Amazon and DHL make use of AMRs equipped with AI for tasks such as material handling, order picking, and warehouse navigation, which are paramount

to the service outcome. These robots are made adaptive to dynamic environments to optimize routes based on AI-generated algorithms.

Practical impact: increased efficiency in warehouse operations, reduced manual labor, and improved order fulfillment processes.

5. Service Robotics:

   Case study: AI-powered social robots

   Application: Softbank's Pepper is a social robot designed to socially interact with humans. This robot uses AI for natural language processing and facial recognition, which enables it to provide information, engage in social conversations, and recognize emotions.

   Practical impact: enhanced customer service in retail settings, interactive assistance in public spaces, and the potential for companionship and support in various contexts and needs.

6. Agricultural Robotics

   Case study: AI-integrated agricultural drones

   Application: drones equipped with AI algorithms are used in precision agriculture applications. These drones can analyze crop health, monitor field conditions, and optimize farming practices based on data-driven insights.

   Practical impact: increased crop yields, reduced resource usage, enhance its ability to address specific areas of concern in large agricultural fields.

7. Search and Rescue Robotics

   Case study: AI-enhanced search and rescue robots

   Application: used for disaster response systems, robots equipped with AI are used to navigate complex terrains, recognize survivors, and assist in search and rescue operations. These robots can adapt to unpredictable conditions given the stability of PID controller AI integrated.

   Practical impact: observed improved efficiency in locating and aiding survivors in disaster-stricken areas, with enhanced safety interaction with rescue teams.

8. Humanoid Robots

   Case Study: AI-driven humanoids in research and assistance

   Application: humanoid robots like Boston Dynamics' Atlas have AI algorithms deployed in their systems for dynamic movement, environmental interaction, and learning from past experiences. They find applications in diverse areas of research, assistance, and search in hazardous environments.

   Practical impact: advancements in human–robot interaction, play potential roles in assistance for the elderly or individuals with some disabilities and exploration of environments unsafe for humans.

9. Environmental Monitoring
   Case study: AI-enabled underwater robots
   Application: underwater robots equipped with AI are essential for environmental monitoring in oceans. These robots can analyze water conditions, track marine life, and contribute to scientific research and equally help in under water military applications.
   Practical impact: improved understanding of marine ecosystems, real-time monitoring of environmental changes, and provide assistance for marine conservation efforts.

These case studies exemplify the diverse and impact-full applications of AI-integrated robotics across various industries. From autonomous vehicles to healthcare, manufacturing, agriculture, and beyond, the synergy of AI and robotics continues to drive efficiency, innovation, and transformation of traditional processes to modern and smarter processes.

3.10.1 Emerging trends and future prospects in AI-integrated robotics

1. Edge AI and On-Device Processing
   Trend: moving AI processing to edge processing for on-board controllers in the robotic system.
   Future prospects: enhanced real-time decision-making, reduced dependence on centralized computing, and improved efficiency in resource-constrained environments.
2. Explainable AI (XAI)
   Trend: growing emphasis on making AI models more transparent and interpretable.
   Future prospects: addressing concerns about the "black-box" nature of advanced algorithms, ensuring accountability, and fostering trust in critical applications.
3. AI for Swarm Robotics
   Trend: applying AI techniques to coordinate and optimize the behavior of swarms of robots.
   Future prospects: distributed decision-making for improved efficiency, scalability in robotic systems, and applications in collaborative tasks such as search and rescue or environmental monitoring.
4. Human–Robot Collaboration Enhancements
   Trend: advancements in harmonizing collaboration between AI-driven robots and human workers.
   Future prospects: enhanced safety, transparency, and guidelines for shared tasks, enabling more seamless cooperation in diverse industries.

5. Continued Advances in Sensing Technologies
   Trend: ongoing developments in sensors, including improved vision systems, tactile sensors, and advanced feedback mechanisms.
   Future prospects: enriching the capabilities of robots for more intricate tasks, better adaptation to dynamic environments, and heightened responsiveness.
6. Ethical Considerations in AI Robotics
   Trend: integration of ethical considerations into AI algorithms for responsible and unbiased decision-making.
   Future prospects: ensuring fairness, transparency, and ethical behavior in robotic systems, particularly in applications where human well-being is a primary concern.
7. AI for Personal Robotics
   Trend: growing applications of AI in personal robotics, including robotic companions and assistants.
   Future prospects: expanding the role of robots in daily life, providing support for individuals with diverse needs, and contributing to a more connected and intelligent living environment.
8. Quantum Computing Impact
   Trend: exploring the potential impact of quantum computing on AI algorithms for robotics.
   Future prospects: unleashing unprecedented computational power, enabling more complex AI models, and potentially revolutionizing the capabilities of robotic systems.
9. Integration of AI with Biotechnology
   Trend: convergence of AI and biotechnology for enhanced robotics in healthcare and life sciences.
   Future prospects: advancements in AI-driven robotic prosthetics, personalized healthcare assistance, and applications in biotechnological research.
10. AI Robotics in Extreme Environments
    Trend: increasing applications of AI-integrated robots in extreme conditions such as space exploration, deep-sea exploration, and hazardous environments.
    Future prospects: robotic systems capable of autonomously navigating and performing tasks in environments that are challenging or inaccessible to humans.

3.11 CONCLUSION

The trajectory toward advancement and improvement of programming platforms for autonomous robots points toward a dynamic and diverse future. The expanding structure of programming paradigms caters to the unique demands of various robotic applications, equally fostering adaptability and efficiency. The rise of middleware and robotics' frameworks coupled with an emphasis on open-source collaboration, is creating a rich ecosystem that encourages research and innovation while enhancing knowledge sharing among developers globally. The integration of simulation environments offers a crucial and intricate bridge between virtual testing and real-world deployment, thus ensuring the optimal, robustness, and safety of autonomous systems. Artificial intelligence and machine learning continue to have crucial roles as integral components of these programming platforms and are poised to empower developers with tools to create innovative, smarter, more adaptive robots. While challenges such as real-time processing, energy efficiency, and security considerations persist, ongoing research and advancements hold the promise of addressing these hurdles. In this landscape, accessibility and education play pivotal roles in democratizing robotic development, which in return ensures that a diverse range of individuals can contribute to and benefit from this evolving field of autonomous robotics. As these platforms continue to evolve, they will not only be shaping the future of robotics but also contributing enormously to a world where intelligent autonomous systems become ubiquitous, transforming industries, factories, diverse sectors, and enhancing our daily lives.

REFERENCES

[1] Thrun, S., & Burgard, W. (2005). "Probabilistic Robotics." MIT Press.

[2] Arkin, R. C. (1998). "Behavior-Based Robotics." MIT Press.

[3] Kortenkamp, D., Bonasso, R., & Murphy, R. (1998). "*Artificial Intelligence and Mobile Robots: Case Studies of Successful Robot Systems.*" MIT Press.

[4] Siciliano, B., & Khatib, O. (2008). "*Springer Handbook of Robotics.*" Springer.

[5] Bagnell, J. A., & Schneider, J. (Eds.). (2014). "*Autonomous Robots: From Biological Inspiration to Implementation and Control.*" Springer.

[6] Quigley, M., Gerkey, B., & S. William D. (2015). "Programming Robots with ROS: A Practical Introduction to the Robot Operating System". O'Reilly.

[7] Hughes, C. & Hughes, T. (2016). "*Robot Programming: A Guide to Controlling Autonomous Robots*". QUE Publishing.

[8] Kumar, V., Ostrowski, M., van den Berg, J., & Ostrowski, J. P. (2018). "*Robotics: A Review of Present and Future Challenges*". Imperial College Press, UK.

[9] Correll, N.(2014). "*Introduction to Autonomous Robots*". MIT Press.

[10] Bordignon, M., Menegatti, E., & Gambardella, L. M . (2012). "*Robotics Middleware: A Comprehensive Literature Survey and Attribute-Based Bibliography*". Wiley Online Library.

Books

1. "*Artificial Intelligence: A Modern Approach*" by Stuart Russell and Peter Norvig: This book is a comprehensive text on the theory and practice of artificial intelligence and includes relevant sections on AI in robotics.

2. "*Robotics and Cognitive Approaches to Spatial Mapping*" by Margaret E. Jefferies and Wai K. Yeap: An insightful book discussing the intersection of cognitive science and robotic navigation.
3. "*The Robotics Revolution and Conflict in the 21st Century*" by P.W. Singer: Singer's work explores the impact of robotics and AI on modern warfare but also offers broader insights into the field of robotics.
4. "*Introduction to Autonomous Robots*" by Nikolaus Correll, Bradley Hayes, and Nikolaos Mavridis: A textbook that covers the fundamental concepts in robotics and how AI is applied to achieve autonomy.

Chapter 4

Emerging trends in mobile robots

Ngum Breayon Kwe and Rashmi Priyadarshini

4.1 INTRODUCTION TO MOBILE ROBOTS

Mobile robots are robotic devices capable of moving and operating in various contexts, alone or with human assistance. Unlike stationary robots that stay in one place, mobile robots can explore and carry out activities in dynamic and often unpredictable environments.

Mobile robots can be small wheeled robots used for logistics and indoor navigation, or large unmanned aerial vehicles (UAVs) and autonomous underwater vehicles (AUVs) used for exploration and surveillance in difficult environments.

4.1.1 Key importance of mobile robots

Technological, scientific and social developments are made possible by mobile robots, essential to many industries and businesses. In manufacturing and warehouse environments, mobile robots automate material handling, assembly and logistics activities to increase efficiency and productivity. Medical staff can achieve better patient outcomes and reduce infection rates when mobile robots participate in patient care, medication distribution, and surgical procedures. Higher yields and sustainable farming methods are achieved in agriculture through the use of mobile robots that optimize crop management, monitor plant health, and automate harvesting processes. Mobile robots play an important role in search and rescue operations and disaster relief by helping to search for survivors, assess damage, and transport supplies to dangerous or inaccessible regions. Mobile robots such as rovers and landers are used in planetary science and space exploration to study celestial bodies, collect data, and conduct experiments to further our understanding of the universe.

4.1.2 Evolution of mobile robots

- *Early Concepts and Prototypes (Pre-20th Century)*: early marvels of technology and ancient mythology gave rise to the idea of robots,

DOI: 10.1201/9781003537083-4

especially mobile robots. However, with the advent of automata and mechanical devices during the Renaissance, the contemporary concept of mobile robots began to take shape. These revolutionary devices laid the foundation for the further development of mobile robots.

- *Industrial Revolution and Early Robotics (19th Century)*: significant advances in automation and technology were made during the industrial revolution. Although most early industrial robots were immobile, this period also saw the introduction of simple mechanical devices capable of moving, such as automated assembly lines and conveyor belts. The mobility of robots in the future will be made possible by these developments.
- *Early Mobile Robot Prototypes (Mid-20th Century)*: the first tests of mobile robots were conducted in the mid-20th century. Scientists and engineers began creating tracked and wheeled vehicles that could be controlled via remote controls and basic systems. Scientific research, remote reconnaissance, and military reconnaissance were the primary uses of these early prototypes.
- *Advancements in Robotics Research (Late 20th Century)*: research in robotics made enormous progress in the late 20th century thanks to advances in electronics, computers, and control systems. More complex movement systems, such as legged robots inspired by living creatures, have been studied by researchers. During this period, sensor technology has also developed, allowing robots to better understand and interact with their environment.
- *Commercialization and Industrial Applications (Late 20th to Early 21st Century)*: the development of microprocessors and smaller sensors has made mobile robots more practical and accessible for use in commercial and industrial environments. To automate material handling and transportation tasks, companies have started introducing mobile robots into factories, warehouses, and logistics operations. This period also saw the emergence of consumer-facing robots, such as robotic toys and vacuum cleaners.
- *Integration of Artificial Intelligence (21st Century)*: Tthe integration of artificial intelligence (AI) and machine learning algorithms with mobile robots during the 21st century marks a key turning point in this industry. With greater autonomy and adaptability, mobile robots can now learn from their mistakes and immediately make wise decisions. In this period, autonomous cars, drones, and robot helpers with advanced artificial intelligence skills have become increasingly common.

- *Emerging Trends and Future Directions*: looking ahead, developments in artificial intelligence, sensor technology, and materials science will continue to drive the rapid evolution of mobile robots.

The field of mobile robotics is changing due to emerging concepts, including bioinspired locomotion, human–robot interaction and swarm robotics. Mobile robots have the potential to completely transform numerous sectors, including healthcare, agriculture, transportation, and exploration, as they become smarter and more adaptable.

4.2 CLASSIFICATION OF MOBILE ROBOTS

Mobile robots can be classified based on a number of criteria, such as their applications, control settings, and mobility methods. Here is a brief overview of the classification of mobile robots

A. *Wheeled robots*

Wheeled robots are robotic systems equipped with wheels or tracks that allow them to move and navigate on flat surfaces. These robots can move forward, backward and sideways using motorized wheels or tracks, allowing them to perform a variety of tasks indoors and outdoors.

Applications

- Wheeled robots are widely used in indoor environments such as warehouses, factories, hospitals, and office buildings for tasks such as material handling, inventory management, and surveillance.
- They are also used in outdoor environments, such as urban environments, construction sites and agricultural areas, for tasks such as inspection, monitoring and maintenance.
- In the consumer market, wheeled robots are used as robot vacuum cleaners, lawn mowers, and personal assistants for home or office tasks.

Advantages

- **Stability:** wheeled robots provide stability and smooth movement on flat, flat surfaces, making them suitable for applications where predictability and precision are essential.
- **Efficiency:** wheeled robots can cover large distances efficiently, making them ideal for tasks that require covering large areas or navigating structured environments.

- **Simplicity:** compared to other locomotion systems such as legs or rotors, wheeled locomotion is relatively easy to implement and control, resulting in cost-effective and reliable robotic platforms.

Challenges

- **Limited terrain accessibility:** wheeled robots may have difficulty traversing rough or uneven terrain, stairs, or obstacles higher than their ground clearance.
- **Maneuverability limitations:** although wheeled robots excel at forward and backward movements, they may have limitations in tight spaces or when precise maneuvering is required.
- **Wheel slip:** under certain conditions such as wet or slippery surfaces, wheeled robots may experience wheel slip, affecting their traction and control.

B. Legged robots

The way mammals, birds and insects move is a source of inspiration for designers of legged robots. Leg-controlled robots strive to achieve comparable levels of dexterity, flexibility, and mobility efficiency by modeling the biomechanics and behavioral patterns of these creatures.

Types of legged robots

- **Quadrupeds:** these robots have four legs, similar to animals such as dogs and cats. Quadruped robots provide stability and versatility, allowing them to navigate a wide range of terrain, including difficult, uneven and confusing environments.
- **Hexapods:** hexapod robots have six legs and resemble insects such as ants and spiders. Hexapods are characterized by their ability to distribute weight and independently adjust the movement of the legs when crossing complex terrain with obstacles.
- **Bipedal and humanoid robots:** bipedal robots have two legs that resemble the human form. Humanoid robots strive to imitate human-like movements and interactions, making them suitable for tasks such as human assistance, entertainment, and human–robot interaction research.

Applications

- **Search and rescue:** legged robots are well suited for search and rescue missions in disaster scenarios where they can navigate through rubble, rubble, and collapsed structures to locate survivors.

- **Exploration:** legged robots are used in exploration missions in environments such as forests, mountains, deserts, and planetary surfaces where wheeled or tracked vehicles may have difficulty maneuvering.
- **Inspection and maintenance:** legged robots can access confined spaces and hazardous environments, making them valuable for inspection and maintenance tasks in industries such as construction, infrastructure, and energy.

Advantages

- **Versatility:** legged robots offer superior adaptability to various terrains and environments compared to wheeled or tracked robots. Their ability to overcome obstacles, stairs, and uneven surfaces makes them invaluable in applications requiring mobility and dexterity.
- **Stability:** despite their complexity, legged robots can maintain stability through advanced control algorithms and feedback mechanisms, allowing them to traverse challenging terrain while maintaining balance.
- **Redundancy:** legged robots have redundancy in their locomotor systems, which allows them to continue working even if one or more legs are damaged or malfunctioning.

Challenges

- **Control and coordination:** achieving stable and efficient locomotion in legged robots requires sophisticated control algorithms to coordinate limb movements, adapt to uneven terrain, and maintain balance.
- **Power and energy efficiency:** legged robots consume a significant amount of energy to drive actuators and on-board electronics, which poses challenges in terms of energy efficiency and battery life.
- **Cost and complexity:** designing and building legged robots can be expensive and technically challenging due to the complicated mechanical and control systems needed to mimic biological locomotion.

C. *Aerial Robots*

Drones, often called flying robots, are unmanned aerial vehicles (UAVs) capable of flight. Drones are adaptable platforms that can be used for deliveries, mapping, photography, aerial surveillance, agriculture, and recreation.

Types of flying robots

- **Multi-rotor drones:** since multiple rotors (usually four or more) are arranged symmetrically, multi-rotor drones are one of the

most popular types of flying robots. They are suitable for a variety of applications, such as mapping, photography, videography, inspection, and recreational use because they provide stability, maneuverability, and adaptability.
- **Fixed-wing drones:** fixed-wing drones are similar to traditional aircraft and rely on them for lift and propulsion. Compared to multirotor drones, they are more effective in terms of endurance and range, making them suitable for tasks such as long-range reconnaissance, agriculture, animal tracking, and aerial surveys.
- **Hybrid and vertical take-off and landing (VTOL) drones:** the benefits of vertical take-off and landing (VTOL) along with increased endurance and range are provided by hybrid drones, which integrate fixed-wing and multi-rotor design elements. They are employed for tasks requiring both gliding and long-range flying capabilities, such as aerial mapping, surveillance and cargo delivery.

Applications

- **Aerial photography and videography:** drones with high-resolution cameras and gimbal stabilization systems are commonly used in the film, real estate, tourism, and advertising industries for aerial photography and cinematography.
- **Mapping and surveying:** aerial robots are useful tools for producing 3D models, topographic surveys, and high-resolution maps of land, structures, infrastructure, and agricultural areas. They deal with construction, land management, urban planning, and environmental monitoring.
- **Infrastructure inspection:** drones with specialized sensors, such as Light Detection and Ranging (LiDAR) scanners and thermal imaging cameras, are used to inspect infrastructure structures such as towers, pipelines, power lines, and bridges. They can quickly and safely detect leaks, structural defects, and other irregularities.
- **Emergency response and disaster management:** aerial robots are critical in emergency response and disaster management as they provide situational awareness, search and rescue assistance, and damage assessment in regions affected by natural disasters, accidents or humanitarian crises.
- **Precision agriculture:** precision agriculture uses drones with cameras and multispectral sensors to monitor crops, detect pests, estimate production, and regulate irrigation. They help farmers optimize inputs, reduce costs, and increase crop yields and quality.

Benefits

- **Mobility and accessibility:** flying robots provide unprecedented mobility and access to remote, dangerous, or hard-to-reach locations, allowing them to collect data and perform tasks that would be difficult or dangerous for humans or animals, or robots on the ground.
- **Cost effectiveness:** drones are cost-effective tools for various applications and offer rapid deployment, flexibility, and scalability compared to traditional manned aircraft or ground-based approaches.
- **Safety:** flying robots increase safety by reducing the need for human operators to enter dangerous environments or perform risky tasks such as climbing tall structures or navigating hazardous terrain.
- **Efficiency and productivity:** drones can cover large areas quickly and efficiently, providing timely and accurate information for decision-making in sectors such as agriculture, construction, infrastructure, and emergency response.

Challenges

- **Regulatory compliance:** drones must comply with aviation standards for safety, security, and privacy when operating in airspace. Drone operators may face obstacles in complying with airspace regulations, licensing requirements, and flight restrictions.
- **Limited endurance and payload capacity:** drones often have limited flight time and payload capacity due to factors such as battery life, weight, and aerodynamic design. This limits their applicability for long duration flights or missions with large payloads.
- **Safety issues:** flying robots can pose a safety risk if misused. Concerns include collisions with manned aircraft, unauthorized surveillance, data breaches, and the possibility of accidents or incidents resulting in property damage or injury.
- **Environmental impact:** the growing adoption of drones raises environmental concerns, including noise pollution, animal harassment, and resource consumption (batteries and materials needed for drone production).

D. *Marine robots*

Marine robots, also called underwater robots or autonomous underwater vehicles (AUVs), are a subset of mobile robots that operate in bodies of water such as oceans, seas, lakes, and rivers. Here is an explanation about marine robots:

Types of marine robots

- **Autonomous underwater vehicles (AUVs):** AUVs are autonomous vehicles with sensors, navigation systems, and communication capabilities. They operate independently and without human intervention and are intended to carry out a wide range of activities, including oceanographic research, underwater mapping, environmental monitoring, and marine exploration.
- **Remotely operated vehicles (ROV):** ROVs are tethered underwater vehicles that operate from the surface. They are equipped with cameras, manipulator arms, sonars, and other equipment to carry out underwater inspections, maintenance operations, rescue, and scientific research.
- **Gliders and profilers:** gliders and profilers are AUVs specially designed for oceanographic research and monitoring. They navigate the water column by changing their buoyancy and flying horizontally using wings or flippers. Gliders and profilers collect data on ocean temperature, salinity, currents, and other variables to study ocean dynamics and climate change.
- **Underwater drones:** underwater drones are small, portable AUVs used for recreational, educational, and commercial purposes such as photography, filmmaking, diving, and exploration.

Applications

- **Oceanography and marine research:** marine robots play a crucial role in oceanographic research as they collect data on water properties, marine life, seabed topography, and ocean currents. They contribute to scientific research on climate change, marine biodiversity, habitat mapping, and ecosystem health.
- **Underwater exploration:** marine robots explore unknown places in the ocean, such as deep trenches, hydrothermal vents, and underwater volcanoes, to discover new species and geological features. They provide information about the Earth's geology, oceanography, and biogeochemical cycles.
- **Environmental monitoring:** marine robots monitor pollution, oil spills, coral reefs, and coastal ecosystems to assess environmental health and identify changes over time. They support nature conservation, habitat restoration, and sustainable management of marine resources.
- **Underwater infrastructure inspection:** ROVs and AUVs inspect underwater infrastructure, including offshore platforms, pipelines, cables and wrecks, for damage, corrosion and structural integrity.

They help prevent accidents, optimize maintenance schedules, and ensure compliance with safety regulations.

- **Defense and security:** marine robots are used for defense and security purposes, including maritime surveillance, underwater reconnaissance, mine countermeasures, and anti-submarine warfare. They improve awareness of maritime space, identify threats, and provide support for military operations in coastal and offshore areas.

Benefits

- **Access to inaccessible environments:** marine robots enable exploration of remote regions in deep and dangerous waters that may be difficult or dangerous for humans. They enable scientists, engineers, and researchers to study underwater ecosystems, geology, and natural events.
- **Autonomy and versatility:** autonomous vehicles (AUVs) can carry out long-duration missions, cover large regions, and adapt to changing environmental conditions without human intervention. They are adaptable platforms that can accommodate large numbers of sensors and equipment for a variety of applications.
- **High-resolution imaging:** marine robots collect high-resolution photos and videos of underwater ecosystems, providing essential data for analysis, documentation, and scientific research. They provide information on marine life, geological structures, and underwater ecosystems that are not visible using typical observation methods.

Challenges

- **Communication and navigation:** underwater communication and navigation are problematic due to radio signal attenuation and poor visibility. Maritime robots use acoustic, optical, and inertial navigation technologies to locate themselves and communicate with humans or base stations.
- **Power and endurance:** maritime robots require efficient power sources and energy management systems to support long-duration missions and energy-intensive sensors. Battery capacity, energy efficiency, and power consumption are essential considerations in the development of AUVs and ROVs.
- **Environmental hazards:** maritime robots require efficient power sources and energy management systems for long-duration missions and energy-intensive sensors. Battery capacity, energy efficiency, and power consumption are important considerations in the development of AUVs and ROVs.

- **Safety and reliability:** ensuring the safety and reliability of marine robots is critical to prevent accidents, equipment failures, and data loss. Robust design, redundant systems, and extensive testing are required to reduce risk and ensure mission success.

E. Hybrid robots

Hybrid robots are an interesting class of mobile robots that incorporate features of different modes of locomotion or combine different types of robotic platforms.

Types of hybrid robots

- **Hybrids with wheels and legs:** these robots offer both the stability and efficiency of wheeled mobility and the agility and adaptability of leg movements. Depending on the environment, they can switch between wheeled and legged modes, allowing them to move on flat ground and navigate around obstacles.
- **Air–ground hybrids:** air–ground hybrids combine the mobility of drones with ground-based capabilities, such as wheeled or legged robots. They can take off and land vertically like drones, then switch to ground mobility for jobs that require greater contact with the environment.
- **Aerial–aquatic hybrids:** hybrid robots combine aerial and underwater movements. These robots can conduct aerial reconnaissance and underwater surveys, providing comprehensive coverage of both terrestrial and aquatic habitats.

Applications

- **Search and rescue operations:** hybrid robots are useful for search and rescue tasks in catastrophic scenarios that require access to remote or dangerous areas. They can travel through wreckage, rubble, and bodies of water to find survivors and help them.
- **Exploration and mapping:** hybrid robots are used to explore and map many areas, including forests, caves, and aquatic habitats. They can collect data, create maps and map landscape features with extreme precision and accuracy.
- **Infrastructure inspection:** hybrid robots inspect infrastructure structures, including bridges, dams, and pipelines. They can enter hard-to-reach areas and perform thorough inspections to reveal structural defects and determine maintenance requirements.
- **Precision agriculture:** precision agriculture uses hybrid robots with sensors and imaging systems to monitor crops, analyze soil, and

detect pests. They provide farmers with vital information on crop health, yield potential, and resource management.

- **Surveillance and security:** hybrid robots improve the security of industrial plants, transit hubs, and public places. They use modern sensor technology to patrol large areas, monitor perimeter borders, and detect suspicious activity.

Advantages

- **Versatility:** hybrid robots use numerous modes of locomotion to navigate different terrains and environments. They can respond to changing situations and perform a variety of tasks with better efficiency and adaptability.
- **Additional capabilities:** hybrid robots can optimize performance in various applications by combining different modes of locomotion that provide stability, agility, speed, and endurance.
- **Cost efficiency:** hybrid robots represent a cost-effective alternative to multimodal mobility, compared to creating and deploying separate platforms for different tasks or situations. They provide a single, modular platform that can meet a wide range of operational requirements.

Challenges

- **Complexity:** designing and controlling hybrid robots can be challenging due to their numerous modes of locomotion and component synchronization. Advanced algorithms and technical solutions are needed to provide smooth transitions between modes and optimize performance in different environments.
- **Energy management:** managing energy consumption and distribution in hybrid robots can be challenging, especially in remote or resource-limited locations. Efficient energy systems and energy harvesting technologies are essential to sustaining long-term missions and increasing operational autonomy.
- **Mechanical design:** hybrid robots must withstand physical forces and environmental conditions during operation. Under harsh conditions, strong mechanical structures, materials, and sealing mechanisms are required to ensure durability, reliability, and longevity.

4.3 EMERGING TRENDS IN MOBILE ROBOTS

Emerging trends in mobile robots summarizes the latest advancements, innovations, and directions shaping the field of mobile robotics. As technology evolves and new challenges emerge, researchers, engineers, and innovators continually seek new solutions to improve the capabilities, performance, and applications of mobile robots. Here are some of the top emerging trends in mobile robots:

- Autonomy and artificial intelligence (AI)
- Collaboration and coordination of several robots
- Human–robot interaction (HRI) and collaboration
- Soft robotics and bio-inspired design
- Mobility and navigation of robots
- Detection and perception of the robot

4.3.1 Artificial intelligence and machine learning

A. Integration of AI in mobile robots

AI and machine learning techniques are seamlessly integrated into many elements of mobile robotics, including vision, navigation, planning, control, and human–robot interaction. AI-based perception algorithms enable robots to understand their environment, identify obstacles, recognize objects, and interpret natural language commands. AI-based navigation and path planning algorithms enable robots to independently navigate complex environments, avoid obstacles, and achieve specified goals efficiently and safely.

Robots using ML-based control algorithms can adjust their behavior and decision-making techniques in response to real-time sensor data and environmental changes. AI and machine learning improve natural language processing, emotion recognition, and social behavior modeling in human–robot interactions, enabling robots to successfully communicate with humans and collaborate seamlessly in a variety of scenarios. •

- Artificial intelligence (AI)

Artificial intelligence refers to the development of algorithms and systems that allow machines, particularly mobile robots, to perform tasks that previously required human intellect. Artificial intelligence enables a variety of tasks in mobile robots, including vision, decision making, planning, and interaction with humans and the environment. Mobile robots can use AI-based approaches to evaluate massive amounts of sensory input from cameras, LiDAR, radar, and other sensors to understand their environment, recognize things, and navigate complex environments. AI-based algorithms enable robots to understand sensory data, make intelligent judgments and perform autonomous tasks, allowing them to function efficiently in dynamic and unpredictable environments.

- Machine learning (ML)

Machine learning (ML) is a subset of artificial intelligence that focuses on creating algorithms and models that allow computers to learn from data and experience without explicit programming. In mobile robotics, machine learning algorithms enable robots to understand patterns, trends, and correlations from sensor data, human interactions and environmental influences. Supervised learning approaches use labeled data sets to teach robots to recognize objects, identify terrain types, and understand human commands. Robots can use unsupervised learning methods such as clustering and dimensionality reduction to discover hidden structures and patterns in data to adapt to new environments and scenarios.

Robots can use reinforcement learning algorithms to learn optimal behaviors by interacting with their environment, receiving feedback on their actions, and modifying their strategy to maximize rewards.

- Progress and opportunities

Rapid advancements in AI and ML technologies, driven by research, innovation, and collaboration in academia and industry, continue to expand the capabilities and applications of mobile robots. Rapid advances in AI and machine learning technologies, driven by research, innovation and collaboration between academia and industry, are expanding the capabilities and applications of mobile robots.

As AI and machine learning algorithms become more powerful and efficient, mobile robots will be able to perform increasingly complex tasks, operate in a variety of contexts, and communicate more intelligently with humans and robots.

The integration of AI and ML in mobile robotics opens new avenues of innovation and disruption in areas such as healthcare, manufacturing, logistics, agriculture and services, where robots, autonomous and intelligent devices can improve productivity, efficiency, and quality of life.

B. Advances in machine learning algorithms

Advances in machine learning algorithms have significantly improved the capabilities and performance of mobile robots. You can find a complete presentation on this topic below::

- Improved perception

Machine learning algorithms have transformed the perceptual capabilities of mobile robots. They help robots better understand sensor data and derive useful information from their environment. Advanced algorithms such as convolutional neural networks (CNN) and recurrent neural networks (RNN) improve object identification, recognition, and tracking, allowing robots to more accurately detect and interact with objects in their environment.

Examples include object detection and recognition, semantic segmentation, gesture and face recognition, terrain classification, object tracking, and anomaly detection.

- Improved navigation and mapping

Mobile robot navigation and mapping relies heavily on machine learning techniques. Algorithms like SLAM (simultaneous localization and mapping) use probabilistic models and optimization techniques to create accurate maps of unknown environments while predicting robot posture. Using reinforcement learning techniques, mobile robots can learn navigation strategies by interacting with their environment, allowing them to find optimal routes and avoid obstacles efficiently.

Examples, simultaneous localization and mapping (SLAM), reinforcement learning for path planning, semantic mapping, predictive modeling for dynamic environments, adaptive path planning and obstacle avoidance, online learning for dynamic environments, etc.

- Control and adaptive behavior

Machine learning allows mobile robots to adapt their control techniques and behavior in response to changes in the environment and task requirements. Reinforcement learning algorithms allow robots to learn from previous experiences and adapt their behavior to maximize rewards, resulting in more resilient and adaptive behavior in dynamic environments.

Examples include the control of autonomous vehicles, the manipulation and grasping of robots, the flight control of quadcopters, the control of robotic arms in manufacturing, the personalized assistance of robots, and the adaptive navigation of mobile robots.

- Human–robot interaction (HRI)

Machine learning algorithms make mobile human–robot interactions more natural and intuitive. Natural language processing (NLP) models help robots understand and respond efficiently to human commands and questions. Sentiment analysis algorithms allow robots to recognize and respond to human emotions, improving the quality of human–robot interaction and collaboration.

Examples include voice assistants and chatbots, social robots for companionship and therapy, personalized recommendations and assistance, gesture recognition and physical interaction, educational and therapeutic applications, assistive robotics for the elderly and disabled, and so on.

- Efficient data processing

Mobile robots produce huge amounts of sensor data, which must be processed effectively to enable real-time decision making. Using machine learning algorithms, robots can quickly analyze and understand sensor data to respond to changing environmental conditions. Bots can use techniques such as online learning and incremental learning to continually modify their models as new data becomes available, ensuring their effectiveness in dynamic contexts.

Examples include real-time decision making, feature extraction and abstraction, adaptive filtering and merging, incremental learning and adaptation, optimized hardware and software architectures, resource-constrained environments, etc.

- Detection and diagnosis of faults

Mobile robots can independently detect and correct errors in their systems and components using machine learning techniques. Anomaly detection systems use sensor data to detect deviations from expected behavior and alert operators to potential problems before they result in catastrophic failures. Examples include anomaly detection in sensor data, predictive maintenance, system health monitoring, fault diagnosis using pattern recognition, and adaptive learning for fault response.

- Continuous learning and improvement

Machine learning allows mobile robots to learn and improve their performance over time by interacting with the environment and receiving input from users. Adaptive learning algorithms allow robots to change their behavior in response to the success or failure of previous actions, ensuring more efficient and effective operation in the long term.

Examples include navigation and map optimization, task learning and optimization, adaptive control and response, error detection and recovery, improving human–robot interaction, and computer optimization.

4.3.2 Swarm robots

Swarm robotics is a paradigm shift in the field of robotics, which emphasizes the coordination and cooperation of large groups of very basic robots

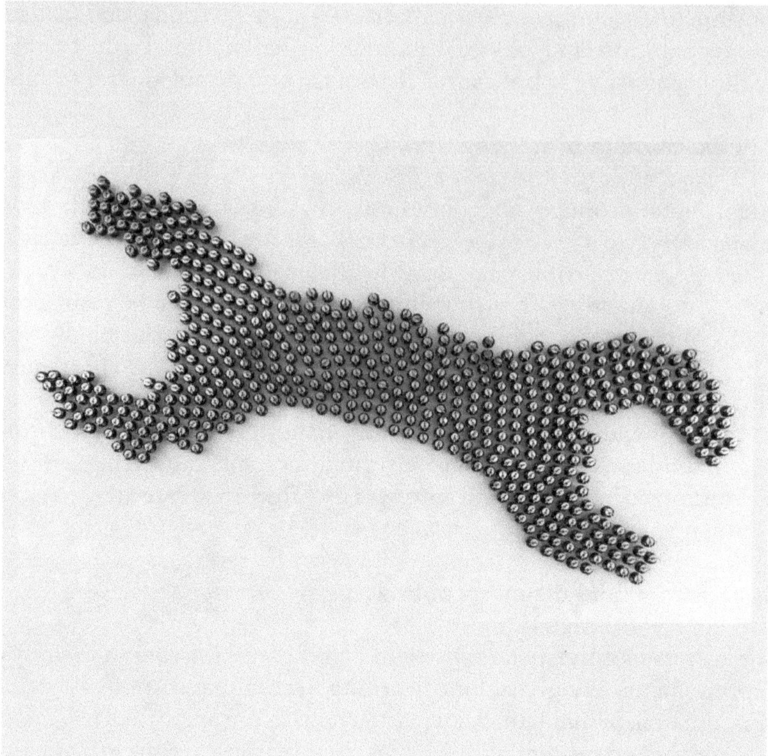

Figure 4.1 Collaboration between swarm robots. (https://golden.com/wiki/Swarm_robotics-AD9D39)

to perform tasks that are beyond the capabilities of individual robots or centralized control systems as shown in Figure 4.1. The idea behind swarm robotics is to perform complex tasks through the collective behavior of several small robots, typically inspired by the behavior of social insects such as ants, bees, and termites.

Here are some key features of swarm robotics:

- Decentralized control and self-organization
  Swarm robotics works on the idea of decentralized control, where individual robots make independent decisions based on local knowledge and ground rules. There is no central command or global strategy; instead, the collective behavior of the swarm is determined by interactions between individuals. Self-organization allows the swarm to adapt to changes in the environment and task requirements

without the need for explicit coordination by a central authority. This decentralized technique improves flexibility, resilience, and scalability in swarm robotics' systems.

- Emergent behaviors and collective intelligence
  One of the distinctive elements of swarm robotics is the creation of complex collective behaviors as a result of individual robot interactions. This emergent behavior often evolves spontaneously as a result of each robot's minimal rules and local interactions with neighboring robots. Collective intelligence is the ability of a swarm to solve problems, make decisions, and perform tasks that would be difficult or impossible for individual robots to perform. The swarm uses the power of collective intelligence to adapt to changing situations and achieve shared goals.
- Robustness and fault tolerance
  Swarm robotic systems are inherently reliable and fault tolerant. Since tasks are distributed across multiple robots, the failure or malfunction of one robot does not necessarily affect the functioning of the entire swarm. The redundancy built into the swarm's robotic systems means that even if some robots are lost or malfunction, the swarm can continue to function and adapt to achieve its goals.
- Scalability and flexibility
  Swarm robotic systems are extremely scalable and allow for variable numbers of robots without requiring major adjustments to the underlying algorithms or coordination methods. This scalability allows swarm robots to undertake tasks at many scales, ranging from small-scale operations to large-scale missions. Furthermore, swarm robotic systems are adaptable to many contexts, tasks, and mission objectives. They can reconfigure and adapt their behavior in response to situational changes and mission demands.
- Applications in various fields
  Swarm robots have various applications in areas such as search and rescue, environmental monitoring, disaster relief, agriculture, surveillance and exploration. In search and rescue scenarios, swarms of robots can effectively survey and map disaster sites, identify survivors, and distribute supplies.
- Inspiration from nature
  Swarm robotics is inspired by the collective behavior observed in social insects such as ants, bees, and termites, as well as other animal groups. By analyzing these natural systems, researchers can discover principles and methods for efficiently designing and controlling artificial swarms.

  Swarm robotics has a wide range of applications in different areas, each benefiting from the different qualities and capabilities of the collective behavior of robots. Here are some applications and advantages of swarm robotics.

- Search and rescue operations
  Swarm robots are very useful for search and rescue missions in dangerous or inaccessible areas, such as fallen buildings, disaster areas, or steep terrain. Swarms of robots can work together to efficiently explore large areas, discover survivors and provide vital information to rescue teams.
- Environmental monitoring and surveillance
  Swarm robotics can be used for environmental monitoring and surveillance tasks, such as collecting data on natural environments, monitoring pollution levels, monitoring wild populations, and detecting changes in the environment. Swarms of robots equipped with sensors and cameras can cover large areas, collect data in real time and provide important information on ecological processes.
- Agriculture and livestock
  Swarm robotics has interesting applications in precision agriculture, where robots can monitor, plant, irrigate, and harvest crops. Swarm robots can optimize resource use, save labor costs, and increase agricultural yields while reducing environmental impact by working together.
- Building and infrastructure
  Swarm robots can assist with site surveys, material transfer, and assembly in construction and infrastructure projects. Robots can work together in difficult conditions to build structures, lay bricks, and perform maintenance tasks by coordinating their activities.
- Exploration and mapping
  Swarm robots are useful for exploratory expeditions in remote or dangerous locations, such as deep sea conditions, caves, and alien surfaces. Swarms of robots equipped with sensors and mapping capabilities can explore unknown terrain, generate accurate maps, and gather important scientific information.
- Distributed sensing and networking
  Swarm robotics enables distributed sensing and networking applications where robots work together to build ad hoc communication networks, monitor environmental variables, and collect data from distant locations. These networks can be used to respond to disasters, monitor wildlife, and monitor the environment.
- Collaborative production
  Swarm robots in manufacturing settings can improve efficiency and flexibility by enabling collaborative manufacturing processes. Robots can collaborate on assembly lines, perform complex tasks, and adapt to changing production needs, resulting in greater productivity and cost efficiency.

- Education and research
  Swarm robotics provides an important educational and research platform in the fields of robotics, artificial intelligence, and collective behavior. Through hands-on experiments and simulations, students, and researchers can learn about topics such as swarm intelligence, distributed control, and emergent behavior.

4.3.3 Human–robot interaction

Human–robot interaction (HRI) is the study, design, and evaluation of interfaces and methods of interaction between humans and robots. It covers a wide range of topics, including the design of intuitive interfaces, the social interaction between humans and robots, and the ethical implications of integrating robots into the human environment as shown in Figure 4.2.
Here is an in-depth look at human–robot interaction:

A. Interface design

Human–robot interaction involves creating interfaces that enable successful communication and interaction. Physical controllers, touchscreens, voice recognition, gestures, and augmented reality displays are examples of interfaces that can be used, depending on the application and circumstances.

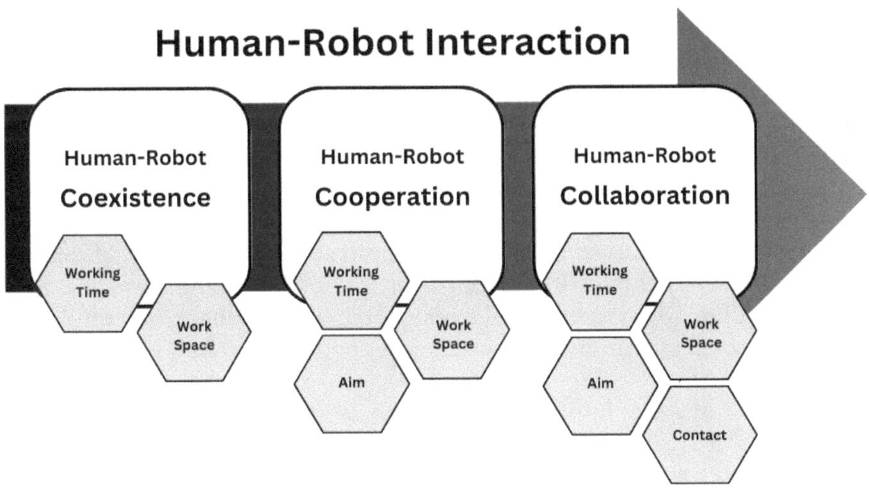

Figure 4.2 Components required in HRI (Schmidtler, Jonas, Knott, Verena, Hoelzel, Christin & Bengler, Klaus. (2015). Human Centered Assistance Applications for the working environment of the future. *Occupational Ergonomics*. 12. 83–95. 10.3233/OER-3.)

B. Natural language processing (NLP)

NLP is key to helping robots understand and respond to human commands and requests. HRI researchers create algorithms for speech recognition, speech understanding, and natural language synthesis so that people and machines can communicate seamlessly.

C. Gesture recognition and body language

By understanding human movements and body language, robots can understand nonverbal signals and communicate socially with humans. HRI research focuses on creating algorithms for gesture identification, posture analysis, and facial expression detection to improve robots' perception and social interaction capabilities.

D. Social robotics

Social robotics is the study of how robots can interact with humans, for example, through communication, collaboration, and emotional expression. Researchers are trying to build socially aware robots that can recognize human emotions, respond appropriately, and develop bonds with humans over time.

E. Collaborative robotics

Collaborative robotics involves humans and robots working together in shared environments to complete tasks. HRI research focuses on safety, trust, and collaboration in human–robot teams, with the goal of creating cooperative environments where humans and robots can complement each other's strengths.

F. User experience (UX) and ease of use

HRI focuses on optimizing the user experience and usability of robotic systems so that people's interactions with robots are intuitive, efficient, and rewarding. This includes interface design, feedback systems, and troubleshooting measures that increase user engagement and trust.

G. Ethical and social implications

HRI research examines the ethical and sociological implications of integrating robots into the human environment, including privacy, autonomy, job displacement, and cultural acceptance. Researchers work with ethicists, policymakers, and others to address these complex issues and ensure the appropriate use of robotic technology.

H. Adaptive and personalized interaction

Personalizing robot behavior based on individual user preferences and modifying interaction techniques based on context are important areas of HRI research. Adaptive interaction models allow bots to learn from human feedback, adjust their behavior over time, and provide personalized assistance based on each user's needs.

Human–robot interaction is a multidisciplinary subject of study that focuses on the design, implementation, and evaluation of interfaces and interaction modalities between humans and robots. HRI aims to promote good and meaningful interactions between humans and robots by fostering transparent communication, collaboration and trust.

4.3.4 Sensor and perception

Sensory awareness and perception are essential elements of robotics, allowing robots to interact with their environment, understand it, and make informed judgments.

Detection

Sensor technology is the ability of a robot to collect information about its environment using various sensors. These sensors can detect various physical properties, including light, sound, temperature, pressure, proximity, and motion.

Robots can be equipped with different types of sensors depending on their intended application and the specific activities they need to perform. Cameras, LiDAR (Light Detection and Ranging), ultrasonic sensors, infrared sensors, encoders and GPS (Global Positioning System) modules are some of the most commonly used sensors in robotics. Sensor data provides robots with important information about their surroundings and helps them identify objects, barriers, landmarks, and other important features.

Perception

Perception is the process by which a robot evaluates and interprets the sensory data it receives from its environment. Perception includes activities such as object recognition, scene understanding, localization, mapping, and tracking. Object identification is the process of recognizing and categorizing objects based on their visual appearance, shape, size, color, and texture. The robots use machine learning algorithms and computer vision techniques to detect and differentiate objects in their environment. Understanding scenes involves analyzing the spatial structure of the environment, recognizing barriers, defining viable paths, and calculating distances to adjacent elements. Robots use techniques such as simultaneous localization and mapping (SLAM) to create maps of their environment and locate themselves within it. Tracking is the process of continuously monitoring the movements of objects or entities over time. Moving objects, people or vehicles can be tracked by robots using motion sensors, cameras or other tracking devices.

In practice, sensing and perception are deeply intertwined in a robot's control and decision-making processes. Sensor data is processed, filtered, and combined to provide a consistent and accurate representation of the environment. Sensor fusion approaches integrate data from different sensors to increase the reliability and resilience of perception systems. For example, data from cameras, LiDAR and inertial sensors can be integrated to gain in-depth insight into the robot's environment.

Sensor noise, ambiguity, occlusions, dynamic environments, and changing lighting conditions all pose challenges for robot sensing and perception. Robust perception algorithms and sensor calibration approaches are needed to address these issues. Recent breakthroughs in sensor and perception technologies, such as deep learning, neural networks, 3D sensing, and probabilistic algorithms, have significantly increased the ability of robots to detect and communicate with complicated situations.

As robotics advances, research into sensing and perception is important to enable robots to function autonomously, travel safely, and perform complex tasks in real-world conditions.

A. Advances in sensor technologies

Advances in sensor technology have played an important role in improving the capabilities of robots in a variety of disciplines, allowing them to perceive, understand, and respond to their environment with better precision and efficiency. Here are some advances in robotic sensor technology:

- Improved detection modes
- Miniaturization and integration
- Multi-modal sensor fusion
- 3D sensing and depth perception
- Advances in LiDAR technology
- Innovations in inertial sensors
- Emergence of environmental sensors
- Integration with AI and machine learning

Overall, developments in sensor technology have had a significant impact on the capabilities and acceptance of robotic systems in a wide range of applications, including industrial automation and logistics, healthcare, agriculture and cities. Continued research in sensor technology is expected to increase the capabilities of robotics to address challenging situations in the real world.

B. Enhanced perception capabilities

Enhanced perceptual capabilities in robotics refer to advances and improvements in the way robots see and interpret their environment. This upgrade includes various elements to detect, process, and interpret sensor data. The following is an exhibit on enhanced perceptual capabilities in robotics.

4.3.5 Sensor technologies

Technological advances in sensors often pave the way for improved perception. Robots observe their surroundings using a range of sensors such as cameras, LiDAR, ultrasonic sensors, radars, infrared sensors, etc. These sensors collect information such as object identification, distance measurement, environmental mapping, etc.

4.3.6 Increased sensory range and precision

As sensor technology advances, robots can sense their surroundings with greater range and accuracy. For example, LiDAR sensors can now identify objects at greater distances and with higher resolution, allowing robots to navigate more complex environments more safely and efficiently.

4.3.7 Multimodal sensor fusion

Improving perception often requires combining information from many senses using multimodal sensor fusion methods. Robots can create richer, more accurate representations of their environment by combining data from many sensor modalities such as vision, depth, and touch sensors. This fusion increases perceptual reliability and robustness.

4.3.8 Machine learning and AI algorithms

Machine learning and artificial intelligence (AI) algorithms play an important role in improving perceptual capabilities. These algorithms allow robots to analyze sensor data, recognize patterns, and extract useful information from their environment. For example, object recognition algorithms can detect and classify objects in a robot's field of view.

4.3.9 Real-time processing and decision-making

Enhanced vision capabilities allow robots to interpret sensor inputs in real time and make autonomous judgments. Robots can adapt to changing situations, avoid obstacles, and perform complex tasks more efficiently and safely by quickly analyzing and interpreting sensory inputs.

4.3.10 Adaptive perception

Some advanced robotic systems have adaptive sensing capabilities, which allow them to learn and improve their sensing capabilities over time. Robots can develop their own perception models, adapt to changing conditions, and improve their performance in a variety of tasks and environments through feedback loops and iterative learning procedures.

4.3.11 Application-specific perception

Enhanced perceptual capabilities are designed for certain applications and use cases. Perception systems, for example, in self-driving cars, are designed to identify and track other vehicles, pedestrians, and traffic signs. Perception systems enable industrial robots to perform precise item handling, assembly, and quality control tasks.

4.3.12 Human–robot interaction

Better perception also leads to better interactions between humans and robots. Robots with high perceptual capabilities can understand human gestures, facial expressions, and voices, enabling more natural and intuitive communication and collaboration with humans.

4.3.13 Autonomous navigation

Autonomous navigation is the ability of a robot or vehicle to navigate and move in its environment without direct human control or intervention. It involves using a variety of sensors, algorithms, and decision-making processes to observe the environment, plan a route, and execute movement commands to reach the desired location. The main features of autonomous navigation are as follows:

- Perception and perception
- Localization and mapping (SLAM)
- Route planning
- Decision making and control
- Adaptation and learning
- Safety and reliability

Autonomous navigation has applications in several areas, including autonomous vehicles, mobile robotics, drones, underwater vehicles, and space exploration. Advances in sensor technology, algorithms, and processing power are driving autonomous navigation innovation, allowing robots to function independently in more complex and dynamic situations.

In the context of autonomous navigation, we will delve into route planning, localization, and mapping (SLAM).

A. Localization and mapping

Localization and mapping, also known as SLAM (simultaneous localization and mapping), is a crucial part of autonomous navigation systems, particularly in robotics and autonomous vehicles.

Location

Localization is the process of determining the position and orientation of a robot in its environment. It provides precise answers to the question "Where am I?" in real time. Robots use a variety of sensor inputs, including GPS, LiDAR, cameras, and inertial measurement units (IMUs), to determine their position relative to a defined reference frame or map.

Localization algorithms often use techniques such as Kalman filters, particle filters (Monte Carlo localization), or probabilistic approaches to combine sensor data and estimate the pose (position and orientation) of the robot relative to the map.

Mapping

Mapping involves creating and updating a representation of the robot's environment. It answers the question, "What does the environment look like?"

Robots use sensor data to build a map of their environment, including obstacles, landmarks, and other relevant features. Commonly used mapping techniques include occupancy grid mapping, feature-based maps, and geometric maps. Mapping algorithms process sensor measurements, extract features, and integrate the information over time to compile an accurate, up-to-date map of the environment.

Simultaneous localization and mapping (SLAM)

SLAM is a computational approach that integrates localization and mapping into a single integrated framework. It allows a robot to estimate its own position while simultaneously creating a map of its environment without prior information. SLAM algorithms iteratively update the pose estimate and robot map using incoming sensor data, accounting for measurement uncertainties and errors.

SLAM algorithms are divided into two types: feature-based SLAM, which detects and tracks distinctive features of the environment, and direct SLAM, which operates directly on raw sensor data without feature extraction.

Challenges and considerations

- Localization and mapping face several challenges, including sensor noise, uncertainty, dynamic environments, occlusions, and perceptual aliasing.

- Robust localization and mapping algorithms must be able to overcome these challenges and adapt to changing conditions to obtain accurate and reliable estimates of the robot's pose and environment.

Furthermore, real-time performance, computational efficiency, and memory limitations are important considerations in the design and implementation of localization and mapping systems for practical robotic applications.

Applications
Localization and mapping are critical components of autonomous systems in many disciplines, including autonomous vehicles, mobile robotics, drones, underwater exploration, and augmented reality. In robotics, precise localization and mapping enable robots to travel freely, design efficient paths, avoid obstacles, and complete tasks in complex, dynamic situations.

B. Path planning and obstacle avoidance

Path planning and obstacle avoidance are essential components of autonomous navigation systems that enable robots and vehicles to move safely and efficiently in dynamic environments.

Path planning
Path planning creates a collision-free path from the robot's current location to its destination, taking into account environmental constraints such as obstacles, topography, and dynamic changes. Path planning algorithms aim to optimize a variety of criteria, such as the shortest path, the fastest path, the most energy-efficient path, and approaches that maximize safety or minimize danger.

Common path planning algorithms include the following:

- **Dijkstra algorithm:** finds the shortest path between nodes in a graph by iteratively selecting the node with the lowest cost.
- **An algorithm (A-star):*** combines the advantages of Dijkstra's algorithm with heuristics to efficiently search for the shortest path.
- **Rapidly exploring random trees (RRT):** constructs a tree of possible paths in the configuration space and explores regions with a higher probability of containing the optimal path.
- **Potential fields:** assigns attractive forces to the target and forces repelling obstacles to guide the robot along a path with minimum potential energy.

Path planning algorithms must take into account kinematics, dynamics, and possible constraints on its movement, such as maximum speed, acceleration, and turning radius.

Obstacle avoidance

Obstacle avoidance involves detecting and avoiding obstacles in the robot's path to avoid collisions and maintain safe navigation. Sensors such as LiDAR, ultrasonic sensors, cameras, and depth sensors are used to detect obstacles and their surroundings. Once obstacles are recognized, the robot must choose the best course of action to avoid them while remaining on the intended path.

Obstacle avoidance strategies may include the below

- **Reactive approaches:** these methods respond in real time to immediate sensory information, such as stopping, slowing down, or changing direction when an obstacle is detected.
- **Predictive approaches:** these methods anticipate potential collisions based on the robot's trajectory and obstacle movement, and proactively adjust the robot's trajectory to avoid future collisions.
- **Combination of reactive and predictive techniques:** many modern navigation systems use a combination of reactive and predictive techniques to achieve robust and reliable obstacle avoidance in dynamic environments.

The effectiveness of obstacle avoidance algorithms depends on factors such as sensor accuracy, latency, environment complexity, and robot speed and maneuverability.

Integration with localization and mapping

- Path planning and obstacle avoidance algorithms are often integrated with localization and mapping (SLAM) systems to provide a complete autonomous navigation solution.
- Localization and mapping provide the spatial awareness and environmental information necessary for path planning and obstacle avoidance algorithms to operate effectively.
- Real-time feedback from path planning and obstacle avoidance helps refine location and mapping information, resulting in more accurate and faster representations of the environment.

Energy efficiency and sustainability:

Energy efficiency and sustainability are key considerations when designing and operating autonomous systems in robots.

Energy efficiency

Energy efficiency is the optimization of energy consumption to reduce waste and maximize performance. Energy savings in autonomous systems are essential to extend operational reach, increase sustainability, and reduce operating costs. The development of energy-efficient hardware components such as low-power processors, sensors, actuators, and communication modules helps reduce overall energy consumption. Software optimization approaches such as efficient algorithms, task scheduling, and energy management strategies can reduce computational effort and operational energy consumption. Solar panels, kinetic energy harvesting systems, and regenerative braking are examples of energy harvesting technologies capable of collecting ambient energy and converting it into usable electrical energy, thereby supplementing energy sources on-board while improving autonomy.

Sustainability

Sustainable development means meeting current needs without compromising the ability of future generations to meet their own needs. In the context of autonomous systems, sustainability encompasses ecological, social, and economic considerations. The goal of environmental sustainability is to reduce the ecological footprint of autonomous systems by minimizing energy consumption, emissions, and resource use. This includes the use of environmentally friendly materials, the reuse of components and the integration of renewable energy sources. Social sustainability requires considering the social impacts of autonomous systems on communities, workers, and society as a whole. This includes considerations such as job relocation, labor practices, equity, accessibility, and social acceptance of autonomous technology.

Economic sustainability refers to the financial viability and long-term viability of autonomous systems. This includes evaluating the total cost of ownership, return on investment, and economic benefits against the risks associated with deploying and operating autonomous systems.

Challenges and solutions

Despite the benefits of energy efficiency and sustainability, there are significant barriers to achieving these goals in autonomous systems. These include technological limitations, government restrictions, market factors, and social attitudes. Addressing these problems requires a comprehensive strategy that includes research and development of innovative technologies, collaboration between industry, academia and government agencies, and the adoption of supporting laws and incentives.

Emerging technologies such as artificial intelligence, the Internet of Things (IoT), and improved materials offer prospects for increasing energy efficiency and sustainability in autonomous systems through intelligent optimization, predictive analytics, and intelligent resource management. Battery technology is key to the development of energy-efficient and sustainable autonomous navigation systems. Furthermore, renewable energy sources are widely used when considering the efficiency and sustainability of robotic navigation systems.

A. Battery technology

Autonomous systems, such as robots, electric cars, drones, and wearable electronics, rely heavily on battery technology. Concerns related to battery technology are as follows:

- **Energy storage capacity:** battery systems differ in their energy storage capacity, which affects the amount of energy that can be stored and delivered per unit weight or volume. Batteries with a higher energy density can provide more energy for longer periods of time, allowing standalone devices to continue operating between charges.
- **Chemistry and composition:** different battery chemistries offer different performance characteristics, such as energy density, cycle life, charging time, and safety. Common battery chemistries include lithium ion (Li-ion), lithium polymer (LiPo), nickel-metal hydride (NiMH), and lead acid. Each chemistry has its advantages and disadvantages and is suitable for specific applications based on factors such as cost, weight, and environmental impact.
- **Fast charging and discharging:** autonomous systems require fast charging capabilities to reduce downtime and increase productivity. Battery technologies that enable rapid charging and discharging enable rapid transitions between missions or operations, increasing overall efficiency and responsiveness.
- **Lifespan and durability:** autonomous systems must evaluate the number of charge-discharge cycles a battery can withstand before drastically reducing its capacity. Batteries with longer life and excellent durability can withstand multiple charge and discharge cycles, resulting in lower maintenance costs and longer life.
- **Safety and reliability:** battery safety is critical, especially in applications where failures or malfunctions can harm people, property, or the environment. Advanced battery management systems (BMS) monitor temperature, voltage and current to prevent overcharging, overdischarging, and thermal overcharging, improving safety and reliability.
- **Temperature sensitivity:** extreme temperatures can reduce battery efficiency, accelerate battery deterioration and pose safety risks. For autonomous systems that operate in a variety of conditions, battery technologies that operate effectively over a wide temperature range or include thermal management systems are desirable.
- **Environmental impact:** sustainable and environmentally friendly battery technologies are becoming increasingly important due to environmental concerns and resource depletion. Manufacturers are looking for alternatives to traditional battery chemistry, such as solid-state batteries, organic batteries and aqueous electrolyte-based batteries, which have greater recyclability, less dependence on rare earth elements and less environmental impact.

- **Integration and form factor:** the size, weight, and form factor of batteries influence the design, integration, and mobility of autonomous systems. Compact, lightweight battery packs with customizable form factors allow for easy integration between platforms while optimizing space and weight distribution.

Battery technologies are evolving to meet the growing needs of autonomous systems in terms of energy consumption, reliability, safety, and sustainability. Advances in battery chemistry, manufacturing methods, and energy management systems are pushing autonomous systems forward and expanding their capabilities across a wide range of applications and industries.

B. Renewable energy sources

Renewable energy sources are natural resources that can be replenished or regenerated in a relatively short period of time, making them more sustainable than nonrenewable energy sources such as fossil fuels. These energy sources are abundant and environmentally beneficial, with the potential to significantly reduce greenhouse gas emissions and dependence on scarce resources. Here are some renewable energy sources:

- Solar energy
  Solar energy harnesses the power of sunlight using photovoltaic (PV) cells, which convert sunlight directly into electricity. Solar panels can be installed on roofs or in fields or integrated into building materials to provide clean, renewable electricity. Solar energy is abundant, available everywhere, and can be used in both sunny and cloudy climates. It is a flexible energy source that can be used residentially, commercially and in public utilities.
- Wind energy
  Wind energy uses wind turbines to convert the kinetic energy of the wind into electricity. Wind turbines are positioned both onshore and offshore in locations with strong and persistent wind patterns. Wind energy is a well-established technology with excellent scalability and low operating costs. It is one of the fastest growing renewable energy sources in the world and helps accelerate the transition to a low-carbon economy.
- Hydroelectricity
  Hydropower uses the energy of flowing water, usually by building dams on rivers or reservoirs to create artificial lakes. The force of the water flow drives turbines connected to generators and producing energy. Hydropower is a robust and well-established renewable

energy source that provides baseload power and system stability. It can also be coupled with other renewable energy sources to help balance supply and demand.

- Biomass energy
  Biomass energy comes from organic resources such as wood, agricultural waste, and organic waste. Biomass can be burned directly to generate heat or electricity, converted to biogas via anaerobic digestion, or transformed into biofuels such as ethanol and biodiesel. Biomass energy is considered renewable as organic resources can be regenerated through environmentally friendly forestry, agriculture, and waste management activities. However, the environmental impact depends on the sustainability of the production of raw materials and processing technologies.
- Geothermal energy
  Geothermal energy converts heat from the Earth's interior into electricity or provides direct heating and cooling. It involves tapping into geothermal reservoirs through wells or drilling and using the natural heat to drive turbines or heat exchangers. Geothermal energy is a stable and reliable source of renewable energy that is always available, regardless of weather conditions. It is particularly suitable for places with strong geothermal potential, such as volcanic zones and tectonic plate boundaries.
- Tidal and wave energy
  Tidal and wave energy uses the kinetic energy of ocean currents and waves to produce energy. Tidal barriers, tidal turbines, and wave energy converters are used in coastal locations to harness energy from the ocean. Tidal and wave energy systems are still in the early stages of their development, but they hold enormous promise for generating clean, predictable energy from the world's oceans.

4.4 APPLICATIONS OF EMERGING MOBILE ROBOT TRENDS

Emerging trends in mobile robots have a wide range of applications, which are developing rapidly thanks to advances in robotics, artificial intelligence, sensor technologies, and mobility platforms. Here are some important uses:

- Industrial automation
- Care and medical assistance
- Farming and farming
- Disaster response and search and rescue
- Surveillance and security

A. Industrial automation

Industrial automation is the use of modern technologies such as robots, artificial intelligence, sensors, and control systems to automate industrial processes and jobs that were previously managed by humans. The goal of industrial automation is to increase efficiency, productivity, quality, and safety while reducing costs and error rates.
Here are some important aspects and components of industrial automation:

- **Robotics:** industrial robots are programmable devices that can perform tasks accurately, quickly, and consistently. They are used in factories, warehouses, and other industrial environments for tasks such as assembly, welding, painting, material handling, and packaging.
- **Sensors and actuators:** sensors detect and measure various characteristics, including temperature, pressure, position, velocity, and force. In contrast, actuators are devices that convert electrical impulses into physical movement or mechanical action. Sensors and actuators work together to create feedback control systems that allow machines and processes to respond to changes in their environment and adjust their behavior accordingly.
- **Programmable logic controllers (PLC):** PLCs are specialized industrial computers that control production processes and machines. They are programmable and can be designed to monitor sensor inputs, make logical decisions, and transmit commands to actuators using established logic and algorithms.
- **SCADA (supervisory control and data acquisition) systems:** SCADA systems provide real-time monitoring, control, and visualization of industrial processes and devices. They collect data from sensors and equipment, present it on graphical interfaces, and allow operators to monitor and manage industrial processes remotely through centralized control rooms.
- **Human machine interface (HMI):** HMIs are graphical user interfaces that allow operators to control industrial machines and processes. They provide operators with simple views, alerts, and status indicators that allow them to monitor performance, identify problems, and make informed decisions.
- **Artificial intelligence (AI) and machine learning:** artificial intelligence technologies, such as machine learning, neural networks, and predictive analytics, are rapidly being integrated into industrial automation systems. Artificial intelligence allows robots to learn from data, discover patterns, optimize processes, and make autonomous decisions without explicit programming.
- **Internet of Things (IoT) and connectivity:** IoT technologies enable the interconnection of devices, sensors, and machines in industrial

environments. By connecting disparate systems and resources, IoT facilitates data sharing, remote monitoring, predictive maintenance, and optimization of industrial processes.

- **Safety systems:** industrial automation systems include safety features and protocols to protect workers, equipment and the environment from accidents and hazards. Safety measures such as emergency stop buttons, safety interlocks, light curtains, and safety scanners are integrated into automation systems to ensure safe operation.

B. Healthcare and medical assistance

Healthcare and medical assistance are critical areas where automation and technology play an important role in improving patient care, improving medical outcomes, and optimizing health care delivery.
Here are some healthcare automation and medical assistance applications:

- Automation of diagnosis
  Automation technologies enable faster and more accurate diagnosis through automated analysis of medical images such as X-rays, MRIs, CT scans, and ultrasounds. AI-based algorithms can analyze medical images, detect abnormalities, and help healthcare professionals interpret results, leading to early detection of diseases and better patient outcomes.
- Electronic health records (EHR) and health information systems
  EHR systems digitize and centralize patient records, enable seamless information sharing among healthcare providers, and improve care coordination. Automation streamlines administrative tasks such as scheduling, billing, coding and claims processing, reducing paperwork, errors, and administrative burden for healthcare professionals.
- Telemedicine and remote monitoring
  Telemedicine platforms use technology to enable remote consultations, virtual visits and remote monitoring of patients' health conditions. Remote monitoring devices and wearable sensors collect real-time health data, such as vital signs, glucose levels and activity levels, allowing healthcare providers to monitor patients' conditions and proactively intervene when necessary.
- Robotic surgery and minimally invasive procedures
  Robotic surgical systems help surgeons perform complex procedures with precision, control, and minimally invasive techniques. Surgical robots offer greater dexterity, 3D visualization, and vibration reduction, resulting in shorter recovery times, fewer complications, and better surgical outcomes for patients.

- Medication management and dispensing
  Automated medication management systems simplify the distribution, administration, and monitoring of medications in healthcare facilities. Automated pharmacy systems ensure accurate drug dosing, reduce medication errors, and improve patient compliance.
- Patient care and assistance
  Robotics and automation technologies support patient care and assistance in hospitals and long-term care facilities. Robot-assisted patient lifting and mobility aids help reduce the risk of injury for healthcare workers and improve patient comfort and safety.
- Clinical decision support systems (CDSS)
  CDSS uses artificial intelligence algorithms and medical knowledge databases to provide evidence-based recommendations and alerts to healthcare providers. CDSS helps physicians make informed decisions about diagnosis, treatment planning, drug selection, and patient management.
- Management of the health supply chain
  Automatizzeria optimizes the management of stocks, the logistics of the supply chain and the maintenance of medical equipment in health care institutions. The automated stock tracking systems ensure the availability of essential medical supplies, reducing stocks and minimizing waste.

C. Agriculture and farming

Agriculture and agriculture have made significant advances in automation and technology, revolutionizing traditional practices and improving productivity, efficiency and sustainability in the agricultural sector.

Here is the role of automation and technology in farming and agriculture:

- Precision agriculture
  Precision agriculture uses advanced technologies such as GPS, drones, sensors and data analytics to optimize agricultural practices and resource management. Automated systems collect data on soil conditions, weather patterns, crop health and growth rates, allowing farmers to make data-driven decisions regarding planting, irrigation, fertilization and pest control.
- Automated agricultural machinery
  Modern agricultural machinery, including tractors, harvesters, planters and sprayers, are equipped with automation features such as GPS guidance, automatic steering and computer controls. Automated agricultural machinery improves accuracy, reduces labor

requirements and improves operational efficiency, allowing farmers to complete their tasks faster and with greater precision.

- Robotics in agriculture
  Agricultural robots, or agribots, perform various tasks on farms, including planting, weeding, pruning, harvesting and sorting. Robotic systems use computer vision, machine learning and robotic arms to identify crops, detect weeds, and perform delicate tasks with precision, reducing the need for manual labor and increasing productivity.
- Intelligent irrigation systems
  Smart irrigation systems monitor soil moisture levels, weather forecasts and crop water needs to optimize irrigation scheduling and water use. Automated irrigation systems deliver water exactly where and when it is needed, reducing water waste, minimizing runoff and promoting efficient water management practices.
- Vertical farming and controlled environment agriculture (CEA)
  Vertical farming and CEA technologies allow the cultivation of crops indoors using hydroponic, aeroponic or aquaponic systems. Automated climate control, LED lighting and nutrient supply systems create optimal growing conditions for plants, maximize yields, minimize resource consumption and reduce environmental impact.
- Data analysis and farm management software
  Farm management software platforms integrate data from a variety of sources, including sensors, machines, weather stations, and historical records, to provide insights and recommendations for farm management. Data analytics tools analyze trends, forecast crop yields, optimize inputs and track performance metrics, enabling farmers to optimize production, mitigate risks, and improve decision-making.
- Supply chain traceability and food safety
  Automated systems track and trace agricultural products from farm to fork, ensuring transparency, accountability and food safety throughout the supply chain. Blockchain technology, IoT sensors, and digital platforms enable real-time monitoring of product origins, handling practices and quality standards, increasing consumer confidence in food products.
- Sustainable agricultural practices
  Automation and technology support sustainable agricultural practices such as conservation tillage, cover crops and integrated pest management. Automated systems optimize resource use, minimize chemical use and reduce environmental impact, thereby promoting the long-term sustainability and resilience of agricultural systems.

D. Disaster response and search and rescue

Disaster response and search and rescue operations are critical areas where automation and technology play a crucial role in saving lives, mitigating damage, and facilitating recovery efforts during and after natural or man-made disasters.

Here is how automation and technology are used in disaster response and search and rescue operations:

- Remote sensing and monitoring
  Remote sensing technologies, such as satellites, drones, and aerial reconnaissance platforms, provide real-time situational awareness by acquiring high-resolution imagery, thermal data, and geographic information. These technologies allow rescuers to assess the extent of damage, identify dangers, and prioritize rescue interventions in disaster-affected areas.
- Communication and connectivity
  Robust communications networks, including satellite communications systems, cellular networks, and mesh networks, enable reliable and resilient communications between responders, command centers, and affected communities. Portable communications devices, such as portable radios, satellite phones, and mobile apps, facilitate coordination, information sharing, and resource allocation in challenging environments.
- Unmanned aerial vehicles (UAVs) and drones
  Drones and drones are used for aerial reconnaissance, search and rescue missions and for damage assessment in areas inaccessible or dangerous to human responders. Equipped with cameras, sensors, and thermal imaging technology, drones can quickly survey large areas, locate survivors, and identify critical damage to infrastructure.
- Robotics and unmanned ground vehicles (UGV)
  Robotics and UGVs are used for exploration of hazardous environments, debris removal, and structural inspections in disaster areas. Robotic systems equipped with cameras, sensors, and manipulator arms can navigate through rubble, tunnels, and collapsed structures to locate survivors, deliver supplies and assess structural integrity.
- Data analysis and predictive modeling
  Data analytics and predictive modeling tools analyze large data sets, including weather forecasts, demographic information, and infrastructure maps, to predict disaster risks and plan effective response strategies. Predictive analytics helps emergency management agencies allocate resources, preposition supplies, and implement evacuation plans based on expected impact and severity.
- Humanitarian aid and aid logistics

Automation technologies optimize logistics and supply chain management for the delivery of humanitarian aid, medical supplies, food and shelter to disaster-affected areas. Automated inventory tracking, warehouse management systems and distribution networks optimize relief operations, reduce response times and ensure timely delivery of essential supplies to those who need them.

- Collaborative platforms and information exchange
  Collaborative platforms and information sharing systems enable multidisciplinary teams, including government agencies, NGOs, and volunteer groups, to coordinate response efforts, share situation updates and collaborate in real time. These platforms facilitate interoperability, standardization and coordination between different stakeholders, thereby improving the overall effectiveness and efficiency of the response.

E. Surveillance and security

Surveillance and security encompasses a wide range of measures and technologies to monitor, protect and protect people, assets and environments from potential threats, criminal activity and unauthorized access. When taking a closer look at surveillance and security, the methods, technologies and applications used in these areas should be described in detail:

- Physical security measures
  Physical security measures include barriers, locks, fences, gates, and security guards to control access to buildings, facilities, and sensitive areas. Access control systems such as key cards, biometric scanners, and smart locks regulate entry and exit points and ensure that only authorized people have access to protected areas.
- Video surveillance systems
  Video surveillance systems use cameras, sensors, and recording devices to monitor and record activities in public spaces, commercial properties and residential areas. Advanced video analytics technologies analyze images in real time, detect anomalies and generate alerts in the event of suspicious behavior or security breaches.
- Intrusion and alarm systems
  Intrusion detection systems use sensors, motion detectors, and perimeter alarms to detect unauthorized entry or intrusion attempts. Alarm systems trigger audible alerts, notifications, or automated responses when security breaches or suspicious activity are detected, allowing for quick intervention by security personnel or law enforcement.
- Biometric identification and authentication
  Biometric identification technologies, including fingerprint scanners, facial recognition systems, and iris scanners, verify the

identity of individuals based on unique physiological or behavioral characteristics. Biometric authentication improves security by providing secure access to sensitive information, systems, and facilities, while minimizing the risk of unauthorized access via stolen passwords or credentials.

- IT security measures
  Cybersecurity measures protect digital assets, networks, and information systems from cyber threats, data breaches, and malicious attacks. Firewalls, encryption protocols, intrusion detection/prevention systems (IDS/IPS), and antivirus software protect against malware, phishing attempts and unauthorized access to sensitive data.
- Aerial surveillance and monitoring drones
  Surveillance drones equipped with cameras, thermal imaging, and GPS capabilities are used for aerial monitoring, reconnaissance, and surveillance in large areas, remote locations and difficult terrain. Aerial monitoring provides real-time information, situational awareness and early warning of security threats or emergencies, enabling proactive responses, and interventions.
- Mobile and remote surveillance systems
  Mobile surveillance systems consist of wearable cameras, sensors, and surveillance equipment that can be deployed in temporary or remote locations to monitor events, crowds, or critical infrastructure. Remote surveillance systems enable remote monitoring and management of security cameras and sensors through centralized control centers or cloud-based platforms, facilitating 24/7 surveillance and response capabilities.
- Integrated security solutions
  Integrated security solutions combine multiple monitoring and security technologies into a unified platform, enabling seamless interoperability, data sharing, and centralized management of security operations. These solutions improve situational awareness, optimize resource allocation, and improve response times by providing complete insight and control of security resources and activities.

These are just a few examples of the various applications of emerging mobile robot trends. As technology continues to evolve and mature, the potential for mobile robots to transform industries, increase efficiency, and improve quality of life will continue to grow.

4.5 CHALLENGES AND FUTURE DIRECTIONS

A. Challenges
   - **Technological complexity:** advances in mobile robots require the integration of complex technologies such as artificial intelligence,

machine learning, computer vision, and sensor fusion. Managing this complexity while maintaining reliability and scalability presents a significant challenge.

- **Interoperability:** interoperability between robotic platforms, communication protocols and sensor systems is essential for effective collaboration and integration in different situations. Standardization activities are necessary to successfully address interoperability issues.
- **Safety and reliability:** mobile robots present safety issues when operating in dynamic and unstructured situations, including overcoming obstacles. To reduce security risks, systems must be robust, fault-tolerant, and fail-safe.
- **Ethical and legal considerations:** as mobile robots become more autonomous, ethical concerns arise regarding decision making, liability, and liability in the event of accidents or unexpected effects. Clear ethical frameworks and legal laws are needed to resolve these issues.

B. Cost and accessibility

Small- and medium-sized businesses as well as university research laboratories face the challenge of cost-effective development, deployment, and maintenance of mobile robot systems. Access to resources, funding, and knowledge is essential to drive innovation and adoption.

C. Future directions

- **Advances in artificial intelligence and machine learning:** continued research in artificial intelligence and machine learning will improve the ability of mobile robots to observe, learn, and adapt to changing situations, leading to greater autonomy and decision making.
- **Sensor and hardware innovations:** advances in sensor technology, such as LiDARs, cameras, inertial sensors, and tactile sensors, will improve the perception, navigation, and interaction capabilities of mobile robots, allowing them to function in diverse and challenging contexts.
- **Human–robot collaboration:** improving human–robot interaction and collaboration is critical for mobile robots to successfully support and complement human activities. Research into intuitive interfaces, natural language processing, and shared autonomy will ensure that humans and robots can work together seamlessly.
- **Autonomous navigation and mapping:** advances in SLAM algorithms, path planning, and obstacle avoidance will enable mobile robots to travel independently in dynamic environments such as indoor spaces, urban regions, and hazardous terrains.

- **Robot swarms and multi-agent systems:** research on robotic swarms and multi-agent systems can facilitate collaborative and distributed operations for tasks such as search and rescue, surveillance, and environmental monitoring. Research into swarm intelligence, coordination algorithms, and collective behavior will open new possibilities for mobile robotics.
- **Humanitarian and societal applications:** the use of mobile robots for humanitarian aid, disaster relief, healthcare, and environmental monitoring is promising. Collaboration between academia, business, and government organizations can promote the development of mobile robotic solutions that solve important social problems while improving human life.

4.6 CONCLUSION

In summary, new advances in the field of mobile robots represent a radical shift towards greater autonomy, intelligence and flexibility of robotic systems. Advances in sensing, mobility, and human–robot interfaces are opening the door to a wide range of applications in areas such as manufacturing, healthcare, logistics, and agriculture. However, these improvements come with major problems such as technological complications, security concerns, ethical issues, and regulatory hurdles. Despite these challenges, the future of mobile robots is extremely promising and requires continuous studies, multidisciplinary collaborations, and proactive policy development to properly realize their potential and ensure their good impact on society.

BIBLIOGRAPHY

Al-Fahaam, H., Davis, S., & Nefti-Meziani, S. (2017). "A review of recent trends in human-robot interaction: Perception, action, and cognition." *International Journal of Social Robotics,* 9(4), 465–480.

Arkin, R. C., Balch, T., & BOTA, M. (1998). "*Behavior-based robotics.*" MIT Press.

Beetz, M., Sturm, J., Civera, J., Maldonado, A., & Pomerleau, F. (2015). "Robots in the real world: Understanding long-term Autonomy." *Robotics & Automation Magazine*, 22(1), 91–103.

Chen, Y. S., & Chiang, C. H. (2011). "Development and classification of mobile robot navigation methods." *Journal of Intelligent & Robotic Systems,* 63(3-4), 417–436.

Du, Z., Wang, Z., & Yu, H. (2018). "A review of classification methods for mobile robots." *Journal of Intelligent & Robotic Systems,* 91(3-4), 521–540.

Dudek, G., & Jenkin, M. (2010). "Mobile robot navigation and mapping: A review." *IEEE Transactions on Robotics*, 26(6), 630–653.

Fox, D., & Burgard, W. (2000). "The dynamic window approach to collision avoidance." *IEEE Robotics & Automation Magazine*, 4(1), 23–33.

Javed, A. Y., Jung, J. K., Lee, J. W., & Park, S. (2017). "A review of autonomous navigation and control systems in unmanned aerial vehicles (UAVs) using localization sensors." *Sensors,* 17(7), 1520.

Kim, J. H., & Cho, K. R. (2007). "A classification method for mobile robots based on environmental information." *IEEE Transactions on Robotics,* 23(3), 560–573.

Kolling, A., Kruse, T., & Röhrbein, F. (2015). "Mobile robots for telepresence, telemanipulation, and augmented reality." *Robotics and Autonomous Systems,* 71, 116–133.

Park, J., Kim, D., & Tomizuka, M. (2017). "Human-robot collaboration in mobile manipulation: A literature review." *IEEE Transactions on Human-Machine Systems,* 47(4), 427–443.

Pires, J. N., Marques, L., & Lima, P. (2007). "A taxonomy of mobile robots." In *Proceedings of the 9th International Workshop on Robot Motion and Control* (RoMoCo'07).

Schmidtler, J. & Knott, V. & Hoelzel, C. & Bengler, K. (2015). Human centered assistance applications for the working environment of the future. *Occupational Ergonomics,* 12, 83–95. 10.3233/OER-150226.

Siadat, A., Biglarbegian, M., & Zhu, W. (2016). "A review on classification and control strategies in mobile robots." *Journal of Intelligent & Robotic Systems,* 83(3), 475–505.

Siegwart, R., Nourbakhsh, I. R., & Scaramuzza, D. (2011). *"Introduction to autonomous mobile robots."* MIT Press.

Thrun, S., Burgard, W., & Fox, D. (2005). "Probabilistic robotics." *Communications of the ACM,* 45(3), 52–57.

Zhou, K., & Fu, Y. (2018). "Research progress and development trend of mobile robots in service applications." *Journal of Robotics,* 2018, 1–14.

Books

1. "*Introduction to Autonomous Robots*" by Nikolaus Correll, Bradley Hayes, et al.
2. "*Mobile Robotics: Mathematics, Models, and Methods*" by Alonzo Kelly and Vladimir Sukhoy.
3. "*Robotics: Modelling, Planning and Control*" by Bruno Siciliano, Lorenzo Sciavicco, et al.
4. "*Probabilistic Robotics*" by Sebastian Thrun, Wolfram Burgard, and Dieter Fox.

Chapter 5

Applications of smart autonomous systems in healthcare

Smitha Shivshankar, Neeraj Makhija, and Premkumar Mathusudhanan

5.1 INTRODUCTION

The healthcare industry is undergoing a transformation due to concerns such as rising costs and the shortage of healthcare experts, and events like COVID-19 have exposed these vulnerabilities. The crisis has thrown light on the need to revamp care systems. Technological innovations are adopted at a slow pace, leading to exhaustion among practitioners struggling to stay updated with vast medical information. However, better outcomes can be achieved by simplifying the healthcare system through technology like artificial intelligence (AI).

AI evolution greatly promotes social development and brings in huge economic benefits [1], optimizing human resources, improving efficiency, reducing labor costs, and generating new job demands [2]. AI research began at the end of the 20th century [3] and has attracted significant interest with the potential of AI to automate many tasks currently requiring human intervention from a variety of fields such as computer vision, speech recognition, and natural language processing [4]. AI's potential is visible in various areas of diagnosis and treatment. There were potential biases and gaps in knowledge as traditionally doctors relied on personal experiences and interactions to enhance their practices, however, with AI, vast sources are available to access through which doctors can get insights and best practices.

5.1.1 Evolution of artificial intelligence

The main aim surrounding AI lies in enabling machines to perform exercises to replicate human or animal intelligence and behavior with the recent research spanning over six major sub-disciplines, including machine learning (ML), robotics, computer vision, automated reasoning, machine perception, and knowledge representations [5], also including a number of techniques and applications such as genetic algorithms, neural networks and pattern recognition [6, 7]. AI can be thought of as a "branch of computing" [5].

 DOI: 10.1201/9781003537083-5

AI enables simulation of human intelligence such as learning, reasoning, and problem solving, using computational models and algorithms [8]. AI aims to achieve to simulate human intellectual activities by employing the approaches using the expression methods from acquiring and analyzing the knowledge [9, 1]. AI analyses large growing amounts of data and takes actions based on what is learnt from the data. There is a long process of development starting from the founding period to knowledge engineering, speech translation and speech recognition, use of Graphic Processing Unit (GPUs) that allowed large-scale access to data [1].

AI enables productivity and performance to be improved in industrial, intellectual, and social applications by overcoming the computational and intellectual limitation of humans [10]. Big data and ML are impacting most aspects of modern life. Personal profiling has gained wide acceptance and value for behavioral understanding [11]. ML and deep learning include algorithms for tasks such as regression, clustering that is trained on data [12]. AI is now integrating into daily life, creating more opportunities for continued innovation across industries and becoming indispensable.

5.1.2 AI in Healthcare

Although the healthcare service demands are significantly increasing, there is a struggle to balance the developments in technology and high patient expectations with service levels and outcomes as there is shortage in healthcare practitioners [8]. Information and communication technology (ICT) facilitate operational efficiency and enhance competitiveness by applying the advanced digital technologies for innovating and creating values across industries [12, 13]. The on-demand healthcare services have enabled remote interaction-based healthcare delivery through remote interactions by leveraging the advancements in wireless technology and smartphones through health tracking apps and search platforms [8]. Hospitals and care providers are deploying technologies such as AI, ML smart sensors, big data, and the Internet of Things (IoT) [12, 13].

AI-supported technology provides more opportunities for innovation for improved care [12]. It is imperative for the health leaders to appreciate AI technologies, apply them, and transform healthcare in clinical decision support, patient management, and drug research. On the other hand, bringing AI to success requires overcoming a series of challenges [8] that includes (1) lack of understanding about what a particular type of AI technology can or cannot do; (2) a lack of clear strategies for integrating different AI technologies into the existing care systems to effectively solve the most pressing problems that health organizations currently face; (3) a shortage of a well-trained workforce for AI implementation; (4) the incompatibility of AI technologies with legacy infrastructure; and (5) a lack of access to good and diverse medical data for training ML algorithms.

5.2 AI TECHNOLOGY

AI is a collection of technologies and has immediate relevance to the healthcare field [14]. AI in medicine can be divided into two subtypes: virtual and physical [15]. Applications such as electronic health record systems for guidance in treatments are classified under the virtual category. Robot-assisted surgeries, intelligent prostheses for physically challenged and elderly care are classified under the physical category. Evidence-based medicine draws insights by developing associations and patterns from existing information database, traditionally obtained using statistical methods. The subsequent section delves into the current AI technologies and their application in healthcare.

5.2.1 AI technology in healthcare

5.2.1.1 Machine learning

ML, a statistical technique, involves identifying patterns and training a predictive model with data inputs and making useful predictions from new data, and almost 63% employ it in their business according to the Deloitte survey of 1,100 US managers, 2018 [14, 16]. ML algorithms need not be explicitly programmed but learn and improve from experience. In healthcare, ML is used in precision medicine that predicts a patient's treatment based on different attributes of patients and treatment context [17].

The most common ML algorithms such as supervised learning, unsupervised learning, reinforcement learning, and deep learning are widely used in other types of AI technologies. Neural networks are a more complex from of ML [18] – a technology that associate inputs with outputs by viewing problems in terms of inputs, outputs, and weights of variables or "features." It is also expanding its reach to radiomics, which detects image features not visible to the naked eye. The combination of the two are primarily used in cancer-related image analysis offering a higher diagnostic precision than the earlier computer-aided detection (CAD) tools [19].

5.2.1.2 Natural language processing (NLP)

NLP, another form of AI, enables computers to interpret and use human language by encompassing applications like speech recognition, text analysis, and translation. It involves two approaches: statistical and semantic. Statistical NLP uses ML, especially deep learning neural networks, and needs a large "corpus" of language data [14]. NLP is predominantly used in clinical documentation and research to create, understand, and classify data. It helps in analyzing clinical notes, creating reports, transcribing patient interactions, facilitating conversational AI, improving diagnostic accuracy, and streamlining clinical processes, thereby improving patient care [20].

5.2.1.3 Rule-based expert systems

Expert systems were the primary AI technology in 1980s, which functions on "if-then" rules and have been extensively used in healthcare for clinical decision support. They expect the human expert to draft rules specific to the knowledge domain. While they are considered effective to a certain limit, they falter when rules conflict or become too numerous, and updating them becomes challenging [14]. Also, changing the rules can be burdensome if there is any significant change in the knowledge area. These systems are slowly being replaced by ML and with approaches based on interpreting data using proprietary medical algorithms.

5.2.1.4 Physical robots

Recently, robots have evolved to work collaboratively with humans and are easily trained through demonstration, with increasing AI capabilities embedded in their operating systems. Surgical robots approved since 2000 in the USA, enhance surgeons' abilities for precise and minimally invasive procedures using robotic surgery.

5.3 AI BENEFITS IN HEALTHCARE

AI deployment in healthcare brings forth commendable benefits and significant challenges in healthcare by offering transformative potential. While improved patient management and streamlined health services are promising, they also poses ethical, transparency, and trust-related issues, as its applications expand in the health sector [21]. Some of the AI benefits in healthcare are listed below.

1. Enhancement of patient management choices and outcomes.
2. Fewer referrals, cost reductions, and time savings.
3. Promotion of healthcare delivery in rural areas, aiding in the recruitment and retention of healthcare professionals there.
4. Contributes to a more globally equitable healthcare system.
5. Enables telemedicine, potentially altering the business model in hospitals.
6. Assists doctors in diagnosis, with future potential for detecting a wider range of diseases.

The available improved machine learning algorithms, increased data access, affordable hardware, and 5G availability enables the AI healthcare to reach $187 billion by 2023, which is currently valued at $11 billion in 2021. AI and ML can lead to industry transformations by quickly analyzing vast health data [22].

5.4 AI APPLICATIONS IN HEALTHCARE

AIs' role in healthcare spans from personalized care to operational efficiencies, transforming patient experience and clinical outcomes. AI is making a significant contribution across the healthcare industry in various areas. This section throws light on the contributions of AI [22–28], which make a significant contribution to the following areas of the healthcare industry including the below:

1. **Personalized health information:** helps patients to make informed decisions by offering personalized health recommendations by analyzing patient data.
2. **Remote monitoring:** AI-powered systems monitor vital signs and reduces in-person visits, alerting healthcare providers to potential issues.
3. **Virtual consultations:** facilitates remote medical care that can benefit those in remote areas or with mobility challenges.
4. **Medication management:** reduces adverse events by analyzing patient prescription histories and vital signs to improve medication management.
5. **Transparency:** increases transparency, helping build trust between patients and providers.
6. **Medical radiology:** aids in analysis of medical images, detecting lesions, predicting disease progression, ensuring image quality and minimizing radiation exposure.
7. **Diagnostic histopathology:** enhances image analysis of tissue samples, automates tissue segmentation, predicts disease progression, ensures quality control, and increases efficiency.
8. **Management of patient complaints:** AI automates complaint management, analyses patient feedback to pinpoint improvement areas and predicts maintenance needs for medical equipment.
9. **Enhancing quality of medical care:** AI detects and prevents errors, offers real-time clinical decision support, facilitates continuous medical education, and promotes quality improvement.

Based on the above benefits, we classify the application of AI in healthcare into four main domains: medical diagnosis, drug discovery, transforming patient experience, and robotic surgery. The healthcare is being invented and rejenuvated through machines that can infer, perceive, learn, and act. The following subsections discuss some applications of AI in the healthcare industry for a clear understanding of the rapidly evolving field.

5.4.1 AI in medical diagnosis

Incomplete medical histories lead to deadly human errors such as errors of omission and commission, the former occurring as a result of action not

Table 5.1 AI in medical diagnosis (data from [23, 25])

Area of application	*AI ability*
Medical diagnosis	Helps care teams to react faster by detecting and notifying, enabling providers to discuss options, provide faster treatments and save lives. Assist in accurate diagnosis in pathology and producing quality images. Diagnose patients with automated system by using data from the electronic medical records and updating patient document automatically. AI chatbots to diagnose and treat illness listening to patient's symptoms and concerns. DL tools streamline radiology diagnoses by analyzing unstructured medical data for better insights. Help in diagnosing potentially deadly blood diseases through ML. Medical imaging AI platform for detecting cancer and cardiovascular issues and provide personalized care.

taken and the later occurring as a result of wrong action [23, 25]. Table 5.1 presents the application of AI in medical diagnosis.

5.4.2 AI in drug discovery

Drug development costs are skyrocketing and the research consuming significant human hours. While the clinical trial costs of each drug costs an average of $1.3 billion, only 10% of it is brought to market [24]. The breakthroughs in technology has enabled pharmaceutical companies to take notice of AI's efficiency, accuracy, and knowledge that can be extended [25, 26]. Table 5.2 presents the application of AI in drug discovery.

5.4.3 AI in transforming patient experience

Seamless patient experience with increased efficiency allows treatment for more patients [25] and positive patient experience yields greater profits while negative experience generates a loss [29, 26]. Table 5.3 presents AI's application in transforming the patient experience.

5.4.4 AI for data management in healthcare

AI application in healthcare stops data hemorrhaging, loss of information, and in turn improving the development of new drugs, preventive medicines, and diagnosis. Big data plays a significant role in healthcare in the areas of cancer research, disease detection, research and development, and health insurance risk assessment [27]. Table 5.4 presents the application of AI in healthcare data management.

Table 5.2 AI in drug discovery (data from [24, 25, [26])

Area of application	*AI ability*
Drug discovery	Developing new medicines in immuno-oncology and neuroscience fields. Harnessing computation chemistry and ML to develop effective cancer treatments. Combining AI, cloud, and quantum physics in drug design and development. Assist in tackling serious diseases (with a sample size between 10 to 20 million screened with results delivered 100 times faster). Find developmental drugs for neuromuscular and neurodegenerative disorders for successfully passing clinical trial. Develop technology for single cell morphology by combining AI and microfluids.

Table 5.3 AI in transforming the patient experience (data from [25, 26, 29])

Area of application	*AI ability*
Transforming patient experience	AI-powered and ML solutions for resource optimization and increased response rates. Reversing chronic conditions and offering a mental health benefit solution to keep mental health in check. Automating repeated tasks in claims and data migration. Assist in solving operational challenges medical documentation tools and automated speech recognition. Improve patient operational flow efficiency, augmenting the prioritization of hospital activity. AI-based interactive symptom checker for up-to-date medical information possible. Generate insights using AI-ML and predictive analysis to patient journey, manage patient data, clinical history, and information on payment.

5.4.5 AI in robotic surgery

Treating medical conditions and streamlining hospital processes is greatly assisted by medical robots thereby allowing healthcare workers to focus more on patients [25, 28]. Robot-assisted surgery or robotic surgery increases precision, flexibility, and control in complex procedures in comparison to traditional techniques. Table 5.5 presents the application of AI in surgery assisted by robots.

Table 5.4 AI in healthcare data management (data from [27]).

Area of application	*AI ability*
Data management	Optimize hospital efficiency by harnessing data, better engage patients, and improve treatment, develop personalized health plans and diagnosing early signs of disease. Use AI to identify patients at-risk and suggest treatment options. Analytical platform that uses portal reporting, standardized neuro biomarkers and ML algorithms to success probability at each stage of drug development. Cancer discovery and treatment using image analysis to connect data points. Automate and predict processes by analyzing data across healthcare systems. Classify symptoms and suggest treatment using big data and AI to analyze health and actions of patients. Streamline workflows by automating administrative tasks and meet particular needs and priorities while maintaining accuracy.

Table 5.5 The application of AI in robotic assisted surgery (data from [25, 28])

Area of application	*AI ability*
Robotic surgery	Virtual reality and AI-enabled robots allowing for minimal invasive operations. Uses AI and robotics to treat cancer tumors with precision focusing only the affected areas. Robotic surgery assistant features cameras, robotic arms, and surgical tools for minimally invasive procedures. Robot to facilitate therapy of the heart, tiny robot enters the chest, navigates locations of the heart and administers therapy under physicians' control. Robots to improve precision and performance during surgical procedures.

5.5 HEALTHCARE DIAGNOSTICS AND AI – A CASE STUDY APPROACH

5.5.1 Case study – Radiscan

The RIS market in Australia is segmented by type, deployment mode, component, and end-user. By type, the market is divided into picture archiving and communication systems (PACS), RIS, and voice recognition systems.

PACS is the largest segment, accounting for the majority of the market share in 2021 due to the increasing demand for PACS solutions for managing radiology images. The deployment is divided into on-premises and cloud-based. The components are segmented into hardware, software, and services with the software segment being the largest segment, accounting for the majority of the market share in 2021 due to the high cost of hardware and the increasing demand for software solutions that can be easily integrated with existing systems. By end-user, the market is segmented into hospitals, clinics, and imaging centers. Hospitals are the largest end-user segment, accounting for the majority of the market share in 2021 due to the high demand for RIS solutions to improve the efficiency of radiology departments.

The growth of the independent RIS market is being driven by the increasing demand for RIS solutions from small- and medium-sized hospitals and clinics unlike corporate RIS being driven by increasing demand from large hospitals and teaching hospitals. Independent RIS solutions are more affordable than those offered by corporate vendors and offer a wider range of features and functionality. The market capture breakdown of independent radiology or imaging providers is 60% dominated by small- and medium-sized businesses offering a wide range of radiology services, and the providers located in rural and regional areas, offering a personalized level of care than their corporate counterparts. The corporate radiology or imaging providers with a market capture of 40%, dominated by large, multinational companies offer a wide range of radiology services, located in metropolitan areas and offer a more comprehensive range of services than their independent counterparts.

The market for radiology or imaging services in Australia is being driven by the increasing demand for radiology services from the aged people, growing chronic diseases, and the increasing use of advanced imaging technologies. The independent radiology or imaging providers' market is expected to grow at a faster rate than the corporate radiology or imaging providers' market due to the increasing demand for personalized care from patients and the increase in the small- and medium-sized businesses entering the market.

5.5.2 Solution

DNA Solutions Australia T/A Scanaptics has been one the first imaging solutions provider that facilitates a suite of products as a service to the imaging and radiology industry in Australia. As industry pioneers, we observed quite a few challenges that could not be resolved with the current solutions in place.

Central to the radiology workflow is the radiology information system (RIS), a comprehensive software solution designed to manage imaging data, patient information, and administrative tasks. However, despite their importance, current RIS implementations face several challenges that hinder efficiency and limit their full potential. To explore the issues plaguing current radiology information systems and therefore the need for innovative solutions to overcome these limitations needs to be addressed, as presented in Table 5.6.

Table 5.6 List of the limitations experienced in the current RIS

Limitation of current system	*Description*
Fragmented data management	Leads to disjointed workflows, inefficiencies in accessing comprehensive medical records, also hindering seamless information exchange between healthcare providers and increased risk of errors and impeding efficient decision making and compromises patient care.
Limited interoperability	The inability to seamlessly exchange data with PACS and Electronic Health Records (HER) prevents the access to comprehensive patient data leading to potential diagnostic delays and inefficiencies in care coordination.
Complex user interfaces	The complex user interfaces of many of the existing RIS impede user adoption and workflow efficiency having to spend significant time navigating convoluted menus, redundant data entry, and searching for information combined with potential errors, decreasing productivity.
Inadequate decision support	Lack of advanced decision support tools in the current RIS limits the capabilities of healthcare professionals. Incorporating AI and ML capabilities within RIS can enhance diagnostic accuracy, aid in detecting abnormalities, and provide decision support to radiologists, improving patient outcomes and reducing reporting turnaround times.
Limited analytics and reporting	Lack of robust analytics and reporting capabilities limiting the extraction of meaningful insights from imaging data crucial for quality improvement, resource management and research. Implementing advanced analytics tools enable performance tracking, identification of workflow bottlenecks, and facilitate research initiatives for improving the overall efficiency of radiology services.
Older technology stack with no room for enhancements	Many RISs have no emphasis on automation and are not patient centric and relied on data collection to store records. Workflows were managed using extended resources, entering and scanning manual worksheets essential for clinical decisions. While this is addressed in later developed systems, some of these were not firmly united, often creating unproductivity to another role in the organization.

It is no doubt that the RIS has significantly transformed the management of medical imaging, however, implementation faces notable challenges. Overcoming the limitations is crucial to improve efficiency, patient care, and research within radiology and pave the way for a more efficient and effective future for RIS. The empowerment of radiologists, enhancing clinical outcomes, and the field of radiology can be transformed by embracing innovative solutions that leverage interoperability standards, AI algorithms, and user-centric design principles. It is imperative for healthcare organizations, technology developers, and industry stakeholders to collaborate and invest in cutting-edge solutions that can unlock the full potential of radiology information systems, ultimately benefiting both healthcare providers and patients alike. However, despite the importance of such systems, DNA Solutions' expert team identified several challenges that hinder the efficiency and potential of current RIS implementation. As such, this study sought to assess these issues currently plaguing radiology information systems to identify innovative solutions to overcome the limitations.

The overall intention of this project is to build a vendor agnostic tool chest that transforms an imaging platform and revolutionizes the way medical imaging is conducted. Examples of specific technical objectives relevant to the experimental activities include the following presented in Table 5.7.

- To develop an AI-based booking system;
- To create an automated radiologist reporting system;
- To develop a secure and scalable cloud-based infrastructure to host the imaging platform and ensure seamless access and storage for healthcare providers.

The specific issues to be addressed are outlined below:

- Fragmented data management;
- Interoperability limitations;
- Complex user interfaces;
- Inadequate decision support tools;
- Limitations in analytics and reporting capabilities; and
- Outdated technology stacks

The success of this project will significantly impact patient care by improving healthcare delivery efficiency, reducing diagnostic errors, and enhancing clinical outcomes. Furthermore, it will empower healthcare providers with advanced tools and insights to make informed decisions and provide personalized care to patients.

The project AIMS to build all this in one vendor agnostic tool chest that transforms the imaging platform and revolutionizes the way medical

Table 5.7 Key objectives of research and development

Objective	*Functionality*
AI-based booking system	A voice interactive booking system using ML algorithms to analyze patient data, to intelligently schedule appointments minimizing patient waiting times, optimize resource allocation, and improve overall operational efficiency.
Automated radiologist reporting	A reporting system assisting radiologists in interpreting medical images and generating accurate reports by integrating various images in one space by leveraging DL techniques aiming to enhance diagnostic accuracy by automating repetitive tasks, reducing turnaround times and alleviating the workload of radiologists and providing decision support, enabling them to focus on complex cases and critical patient care.
Utilizing DICOM structured data	Structured data directly from imaging modalities used to fill the data in the reports without any errors using intelligent rules and DL to present the data in best usable format.
Cloud-based infrastructure	Hosting the imaging platform on the cloud infrastructure, ensuring seamless access and data storage for healthcare providers facilitating collaborative workflows, enable remote access to imaging data, and foster data sharing and research collaboration, ultimately driving advancements in the field of medical imaging without having to install applications locally on their devices.

imaging is conducted. This chapter will highlight the significance of the first technical objective of the project – "AI-based booking system."

5.5.3 AI-based booking system – vEngage.ai – revolutionizing patient engagement in healthcare

5.5.3.1 Introduction to Olivia: A new smart and efficient assistant for healthcare providers

Healthcare providers across various practices such as medical imaging, general practice, cardiology, and dental surgery, have been facing challenges in maintaining effective communication with patients in terms of appointment scheduling and follow-up engagements. The traditional methods were often inconvenient, and lacked integration with practice management software. Olivia is an innovative automated communication suite designed explicitly for healthcare providers designed to simulate human-like conversations and tuned to understand the intricate requirements of a clinical setting.

5.5.3.2 Evolution and versions of Olivia

The validity of the functionality, efficiency, and impact on user experience and system performance was tested using multiple iterations of assessments. To ensure a heuristic outcome, several design iterations were conducted on datasets to evolve the rules' engine and its preference algorithm. Table 5.8 presents the key variables identified in ensuring the solution was being measured while improving workflows.

5.5.3.3 System evaluation

The system was evaluated with different types of testing that includes the following:

1. Unit testing: testing individual modules and processes post coding.
2. Integration testing: testing applied to integrated modules.
3. System tests: conducted using defined scenarios on the system.
4. Acceptance testing: conducted by users at identified beta sites.

Table 5.9 presents the different types of tests applied on the systems for evaluation and validation.

Table 5.8 Key variables measured to improve workflows

Variables	*Measurement criteria*	*Evaluation criteria*
Time to enter (TOE)	Time taken to initiate a contact to the system to make a booking and after excluding that are system dependent to indicate whether the functionality was successful in reducing the time it took for a person to make a booking.	
Word error rate (WER)	A common metric to evaluate accuracy that measures the percentage of incorrectly recognized words to the total words spoken.	Preparation, recognition, comparison, and calculation
Confidence scores	Confidence scores assigned to each recognition result, indicating how confident it is in the correctness of the recognition.	Confidence score generation, analysis, thresholding
Vocabulary coverage	Vocabulary word percentage the system can accurately recognize.	Preparation, recognition, coverage calculation
Error analysis	Conducting error analysis to identify patterns in misrecognized words or phrases and identify areas for improvement in the system.	

Table 5.9 Test scenarios applied on the systems for evaluation and validation

Iterations	*Objective*	*Tests*	*Tests details*	*Test outcomes*
Iteration 1: functionality testing	Ensure that the basic functionalities of the AI-based patient booking system are operational and without critical errors.	*Booking process testing*	*Scenario testing* - Test a range of scenarios, including different medical conditions, specialties, and appointment types and verify that AI recommends appropriate healthcare providers based on the patient's stated needs.	*Version 1.10* Finding best time slot which is logical but not practical. Subsequent changes were made to introduce a flexibility to allow for less efficient time slots allowing for travel and preparation buffers.
		Cancellation and rescheduling	*Test cases with known outcomes* - Created test cases with predetermined outcomes to ensure that the AI consistently matches patients and providers correctly. The rest cases could cover various specialties, time preferences, and provider availability.	*Version 1.12* Reduced number of parameters introduced to test the algorithm for more natural outcomes. Introducing extra parameters had a negative impact against the processing and response time to calculate the best fit.
		Data integrity	*Randomized testing* - Generate random patient profiles with varying medical requirements and preferences and evaluate AI's recommendations against industry standards or expert opinions to verify correctness.	*Version 1.15* Time learning was introduced to allow for limited interval on available resources
		Integration testing	*Edge case testing* - Test extreme cases, such as patients with rare medical conditions or complex needs and verify that AI can handle uncommon scenarios and provide suitable recommendations.	*Version 1.18* Data optimization in parsing algorithm to run less queries and process parallel independent parameters and variables.

(*continued*)

Table 5.9 (Cont.)

Iterations	*Objective*	*Tests*	*Tests details*	*Test outcomes*
Iteration 2: performance and scalability testing	Assess the system's performance under various loads to ensure it can handle real-world usage.	*Load testing*	*Provider availability* - Simulate scenarios wherein providers with limited providers have limited availability due to vacations, conferences, or high patient loads and confirm that AI accounts for provider availability when making recommendations.	*Version 2.1* Neural engine was optimized to run and ignore non-relevant data to correct an expected input. E.g. non-numerical was treated as noise to the numerical values only where expected to improve efficiency and vice versa.
		Scalability testing	*Feedback loop testing* - Incorporate feedback from both patients and healthcare providers and assess whether AI's recommendations align with the preferences and expertise of the providers and meet patient expectations.	*Version 2.2:* Nuance contexts added to create a more sensible dialogue that could derive more meaningful information. "Baby Scan" would mean a pregnancy scan, a CT cardiac study was a CT angiogram
		Stress testing	*Language and communication preferences* - Test scenarios where patients prefer specific languages or communication styles (e.g., sign language for hearing-impaired patients) and confirm that the AI considers language and communication needs when suggesting providers.	*Version 2.4 current* Data parsing architecture refabricated to hold local identifiable data at premise on a temporary service and not encrypted directly to the cloud environment. Once transaction was finalized the scheduling data was sent and deleted from local services.
Iteration 3: user experience testing	Evaluate the system from the user's perspective to ensure it is intuitive and user-friendly.	*Usability testing*	*Long-term care and continuity* - Test situations where patients require ongoing care from the same provider over multiple appointments and verify that the AI can match patients with the same provider for continuity of care.	

		UI testing	*Emergency scenarios* - Simulate urgent or emergency cases where immediate attention is required and confirm that AI can prioritize and recommend appropriate providers in critical situations.
		Accessibility testing	*Comparison with human experts* - Have experienced healthcare professionals manually match patients with providers and compare AI's recommendations with the choices of these experts to identify any discrepancies.
Iteration 4: accuracy and reliability testing	*Verify the AI's accuracy in scheduling appointments and handling various scenarios*	*Appointment matching*	*A/B testing* - Implement A/B testing by having the AI recommend providers for some appointments and human staff for others and analyze the success rates of AI-recommended appointments compared to manually matched ones.
		Conflict resolution	*Dynamic factors* - Tested how the AI handles changing factors like provider schedules, patient preferences, and appointment cancellations and assess if the AI can dynamically adjust recommendations in response to real-time changes.
		Error handling	
Iteration 5: integration and data security testing	Ensure proper integration with other systems and safeguard patient data.	*EHR integration testing*	
		Data privacy	

(*continued*)

Table 5.9 (*Cont.*)

Iterations	*Objective*	*Tests*	*Tests details*	*Test outcomes*
Iteration 6: real world testing	Test the system in a real-world environment with actual users to identify issues not covered during previous iterations.	*Beta testing* *Feedback collection*		
Iteration 7: continuous improvement	Establish ongoing monitoring and improvement mechanisms	*Monitoring and analytics* *Regular updates*		

5.5.3.4 Results, discussion, and conclusion

Olivia, as an automated communication suite has significantly revolutionized patient engagement in healthcare. The suite has bridged the gap between patients and healthcare providers, by introducing human-like conversations and ensuring convenience and integration. Coupled with the vEngage platform, the system offers a comprehensive solution promoting better patient outcomes and reduces complications. Olivia stands as a testament to the potential of technology in enhancing patient care. The system has improved the patient booking time by a reduced wait time by at least 80% and the turnaround times for users by at least 72%. The system response was being improved overall to reach a defined target of less than 20 ms per record on the browser, currently the response time was 2 sec on average. The work requires automating physician workflows.

5.5.3.5 Features and integration

1. **Versatility across practices:** Olivia caters to a broad spectrum of healthcare settings, including medical imaging practices, general practice, cardiology practice, and dental surgery.
2. **Convenience at its core:** patients can schedule or reschedule their appointments anytime, choosing a channel that is more convenient for them, which removes the constraints of standard office hours and provides flexibility.
3. **Seamless integration:** by integrating effectively with leading practice management software, Olivia ensures that the patients have a smooth experience when selecting slots and receiving appointment confirmations, eliminating potential glitches and miscommunications.
4. **vEngage platform:** augments the suit that allows healthcare providers to maintain engagement with patients during their recovery at home. The holistic approach encompasses symptom monitoring, ensuring medication, diet, and exercise compliance, offering health tips, and alerting on early warning signs.

5.5.3.6 Benefits

1. **Complete visibility of patient during recovery:** Olivia, combined with the vEngage platform, assures the healthcare providers an uninterrupted view of the patient's recovery process enabling timely interventions and offering a more personalized patient experience.
2. **Patient reported outcome measures:** the healthcare professionals can make informed decisions and tweaks in the treatment process using direct feedback from the patients and real-time monitoring.

3. **Reduced complications and readmissions:** Olivia aids in identifying potential problems earlier by actively monitoring and engaging patients during their recovery and hence reducing complications and the associated readmissions.
4. **User-friendly interface:** Olivia is designed to be easy to use, despite its sophisticated backend operations, ensuring that both patients and healthcare providers can navigate the problem without any hiccups.

5.6 ISSUES AND CHALLENGES OF AI IN HEALTHCARE

The transformative potential of AI is evident, leading to a new era of diagnostics, treatment, and patient care. While there are many promises of AI-driven innovations, it is imperative to understand and navigate the multifaceted challenges they bring forward. The AI integration into healthcare requires scrutiny from the viewpoint of ethical dilemmas, data biases to transparency concerns and trust barriers.

Despite the tremendous developments of AI to improve the treatment process, AI casts various challenges such as ethical dilemmas, privacy and data protection, informed consent, social gaps, medical consultation, empathy, and sympathy. According to [30], autonomy, beneficence, non-maleficence, and justice are four significant medical ethical principles and it needs to be considered before integrating AI with the healthcare system. This section delves into the intricate issues and challenges posed by AI in healthcare and emphasizes the importance of a balanced approach that aligns technological advancements with patient welfare and industry standards [21, 30–33].

1. **Data bias:** training of AI systems using non-representative may lead to biases both that include both human-induced biases and data-induced biases leading to underestimated or overestimated patient risks leading to health disparities [31]. AI must be generalizable and transferable and there is a need to follow data representativeness and diversity (training algorithms from more diverse datasets) and should be equally applicable to all patients.
2. **Security and privacy issues:** errors and flaws in AI algorithms will lead to diagnostic errors, leading to wrong treatments and harming human lives. Secondly, AI algorithms can lead to clinical risks as the performance of the algorithm will worsen when training AI with real world data different from the training datasets. Thirdly, the potential autonomous functions of AI applications need continuous updates, checked and regulated, otherwise may lead to potential risks. The need for strong privacy measures and informed consent is imperative as healthcare data is sensitive. The current laws are insufficient to protect data since the data collected by robots are prone to

malicious attacks, there is wide use of patient data for marketing and sales gathered by social networks, and patient data could be sold to pharmaceutical and biotechnology companies.

3. **Ethical considerations:** AI replacing certain tasks poses a dilemma with respect to ethical responsibilities. Humans should be responsible for overcoming the gaps that exist between machine's rule-based conditions, probabilistic statistics, and human judgement based on emotion as AI is still a tool.
4. **Transparency:** deep learning models although effective, challenge the transparency in the medical field as they are hard to interpret. Transparency is prevalent in applications that uses large datasets to learn functional associations. Hidden trade secrets, algorithm complexity, and inability to understand programming and algorithms are possible reasons for opacity [31].
5. **Trust:** as AI algorithms assign weight to certain factors over others, they are susceptible to discrimination, introducing data bias risks. Interpretability also impacts the trust relationship in AI. Although it does not impact the levels of trust on technology, however, it has a significant influence on users' reliance on AI and healthcare decisions [33].
6. **Accountability:** this is a serious concern as it involves multiple stakeholders, from software developers to healthcare providers. Clinicians need to account for their actions due to the regulatory enforced professional requirement and hence the need to provide safe and good care, while technologists should employ an ethical code of practice. Clinicians need to be personally professionally accountable, while technologists contribute to clinical decision making, the question is to what extent technologists should be accountable [34].
7. **Ethical principles:** when applying AI in healthcare, principles of autonomy, benefit, and non-harm must be adhered. Based on the principles of autonomy, patients can ask questions, need to be aware of the treatment process, and know who takes responsibility if the robots fail. On the other hand, it is imperative to have an in-depth understanding of ethics as AI leads to various social complications. The lack of transparency, reduced productivity levels, and AI biases can result in unethical business practices [35].
8. **Social gaps and justice:** AI increases the threat of social inequality and less social justice with every development leaving a fear of losing jobs with the increase of automated systems, surgical robots, and robotic nurses.
9. **Medical consultation, empathy, and sympathy:** robots miss the empathetic and compassionate environment, which has a significant effect on the healing process.

5.7 CONCLUSION

AI's integration in healthcare paints a future filled with possibilities starting from enhanced patient care to efficient administrative processes. Nevertheless, to harness the complete potential of AI in health service, ensuring a balance between technological advancement and ethical integrity, trust and transparency will be of paramount importance. To ensure that the negative aspects should not outweigh the benefits, it is important to consider humanity and ethics.

This chapter explores the potential of AI and autonomous systems in the transformation of the healthcare domain. The chapter begins by delving into the core AI technologies, such as ML, NLP, computer vision, and robotics that underpin the innovations and serve as the building blocks for the changes. The core of the exploration is centered on the AI applications in healthcare from enhancing medical images, predicting diseases from symptoms, revolutionizing patient care, reducing costs, and improving overall health outcomes.

The chapter is supported with a compelling case study demonstrating the practicality and efficacy of AI in the form of an automated booking system, streamlining the operations of healthcare and augmenting the patient experience. The measurement metrics, test cases, and outcomes of the ongoing project substantiate the tangible benefits that are offered by the technology.

The chapter also sheds light on the challenges, ethical dilemmas, and data privacy concerns that arise with the adoption of AI. Regulatory hurdles and the requirement for healthcare professionals to adapt to these new technologies are significant roadblocks to navigate as AI in healthcare is a continuously evolving field and the possibilities are boundless.

In conclusion, the integration of AI and autonomous systems is pivotal in the evolution of the healthcare industry with the potential to provide more personalized, efficient, and effective care eventually leading to better health outcomes after overcoming the challenges to promote AI adoption.

REFERENCES

[1] Lu, Y. and Xu, L. D. "Internet of things (IOT) cybersecurity research: A review of current research topics", *IEEE Internet of Things Journal*, vol. 6, no. 2, pp. 2103–2115, 2019. doi:10.1109/jiot.2018.2869847

[2] C. Zhang and Y. Lu, "Study on Artificial Intelligence: The state of the art and future prospects," *Journal of Industrial Information Integration*, vol. 23, p. 100224, 2021. doi:10.1016/j.jii.2021.100224

[3] L. D. Xu, "Index," *Expert Systems with Applications*, vol. 16, no. 4, p. 407, 1999. doi:10.1016/s0957-4174(99)00022-6

[4] F. Wang and A. Preininger, "AI in health: State of the art, challenges, and future directions," *Yearbook of Medical Informatics*, vol. 28, no. 1, pp. 16–26, 2019. doi:10.1055/s-0039-1677908

[5] E. Farrow, "To augment human capacity—artificial intelligence evolution through causal layered analysis," *Futures*, vol. 108, pp. 61–71, 2019. doi:10.1016/j.futures.2019.02.022

[6] M. H. Jarrahi, "Artificial Intelligence and the future of work: Human-AI symbiosis in organizational decision making," *Business Horizons*, vol. 61, no. 4, pp. 577–586, 2018. doi:10.1016/j.bushor.2018.03.007

[7] P. Esmaeilzadeh, "Use of AI-based tools for healthcare purposes: A survey study from consumers' perspectives," *BMC Medical Informatics and Decision Making*, vol. 20, no. 1, 2020. doi:10.1186/s12911-020-01191-1

[8] M. Chen and M. Decary, "Artificial Intelligence in healthcare: An essential guide for health leaders," *Healthcare Management Forum*, vol. 33, no. 1, pp. 10–18, 2019. doi:10.1177/0840470419873123

[9] N. Duan, L.-Z. Liu, X.-J. Yu, Q. Li, and S.-C. Yeh, "Classification of multichannel surface-electromyography signals based on Convolutional Neural Networks," *Journal of Industrial Information Integration*, vol. 15, pp. 201–206, 2019. doi:10.1016/j.jii.2018.09.001

[10] D. Gursoy, O. H. Chi, L. Lu, and R. Nunkoo, "Consumers acceptance of artificially intelligent (AI) device use in service delivery," *International Journal of Information Management*, vol. 49, pp. 157–169, 2019. doi:10.1016/j.ijinfomgt.2019.03.008

[11] A. Bohr and K. Memarzadeh, "The rise of Artificial Intelligence in healthcare applications," *Artificial Intelligence in Healthcare*, pp. 25–60, 2020. doi:10.1016/b978-0-12-818438-7.00002-2

[12] D. Lee and S. N. Yoon, "Application of artificial intelligence-based technologies in the healthcare industry: Opportunities and challenges," *International Journal of Environmental Research and Public Health*, vol. 18, no. 1, p. 271, 2021. doi:10.3390/ijerph18010271

[13] S. M. Lee and S. Lim, *Living Innovation: From Value Creation to the Greater Good*. Bingley, UK: Emerald Publishing Limited, 2018.

[14] T. Davenport and R. Kalakota, "The potential for artificial intelligence in Healthcare," *Future Healthcare Journal*, vol. 6, no. 2, pp. 94–98, 2019. doi:10.7861/futurehosp.6-2-94

[15] P. Malik et al., "Overview of artificial intelligence in medicine." *Journal of Family Medicine and Primary Care,* vol. 8, no. 7, p. 2328, 2019.

[16] Deloitte. "*State of AI in the Enterprise.*", 2nd Edition, deloitte.com. www2.deloitte.com/content/dam/insights/us/articles/4780_State-of-AI-in-the-enterprise/DI_State-of-AI-in-the-enterprise-2nd-ed.pdf (accessed Oct. 1, 2023).

[17] S. I. Lee et al., "A machine learning approach to integrate big data for precision medicine in acute myeloid leukemia." *Nature Communications,* vol. 9, no. 1, p. 42, 2018. https://doi.org/10.1038/s41467-017-02465-5

[18] M. Sordoa. "Introduction to neural networks in healthcare." *Open Clinical: Knowledge Management for Medical Care,* 2002.

[19] A. Vial et al., "The role of deep learning and radiomic feature extraction in cancer-specific predictive modelling: a review." *Translational Cancer Research,* vol. 7, no. 3, pp. 803–816, 2018.

[20] S. Barth. "AI in Healthcare" Foresee Medical. Foreseemed.com. www.foreseemed.com/artificial-intelligence-in-healthcare (accessed Oct 10, 2023).

[21] S. Sri et al., "Artificial intelligence in healthcare: Opportunities and risk for future." *Gaceta Sanitaria,* 35, S67–S70, 2021.

[22] IBM Education. "The Benefits of AI in Healthcare." ibm.com, www.ibm.com/blog/the-benefits-of-ai-in-healthcare (accessed Oct. 15, 2023).

[23] T. L. Rodziewicz et al., "Medical error reduction and prevention." 2018.

[24] Cbinsights, "The Evolution of Clinical Trials: How AI, Big Tech, & Covid-19 Could Make Drug Development Cheaper, Faster, & More Effective". Cbinsights.com. www.cbinsights.com/research/clinical-trials-ai-tech-disruption/ (accessed Oct .15, 2023)

[25] R. Velazquez, "Artificial Intelligence in Healthcare: 39 Examples Improving the Future of Medicine". Available at https://builtin.com/artificial-intelligence/artificial-intelligence-healthcare (accessed Nov. 2, 2023).

[26] S. M. Yousef. "Applications of Artificial Intelligence (AI) in healthcare: A review." *ScienceOpen Preprints,* 2021.

[27] A. Schroer, "20 Big Data in Healthcare Examples and Applications." builtin.com. Feb. 09, 2023 https://builtin.com/big-data/big-data-in-healthcare (accessed Nov. 2, 2023).

[28] U. Matthew, "Medical Robots Transforming Healthcare." builtin.com. Apr. 26, 2023. https://builtin.com/robotics/surgical-medical-healthcare-robotics-companies (accessed Nov. 2, 2023).

[29] M. Blackman, "The link between financial success and patient satisfaction." *Medical Economy Journal,* vol. 98, no. 7, 2021. www.medicaleconomics.com/view/the-link-between-financial-success-and-patient-satisfaction

[30] D. D. Farhud, and S. Zokaei. "Ethical issues of artificial intelligence in medicine and healthcare." *Iranian Journal of Public Health,* vol. 50, no. 11, 2021.

[31] J. Zhang, and Z. M. Zhang. "Ethics and governance of trustworthy medical artificial intelligence." *BMC Medical Informatics and Decision Making* 23, no. 1, p. 7. 2023.

[32] F. Gille et al., "What we talk about when we talk about trust: Theory of trust for AI in healthcare." *Intelligence-Based Medicine* 1, 2020, p. 100001. https://doi.org/10.1016/j.ibmed.2020.100001.

[33] V. Tucci et al., "Factors influencing trust in medical artificial intelligence for healthcare professionals: A narrative review." *Journal of Medical Artificial Intelligence,* vol. 5, 2022.

[34] H. Smith, "Clinical AI: Opacity, accountability, responsibility and liability." *AI & Society,* vol. 36, no. 2, pp. 535–545, 2021.

[35] H. O. Khogali, and S. Mekid. "The blended future of automation and AI: Examining some long-term societal and ethical impact features." *Technology in Society,* vol. 73, p. 102232, 2023.

Chapter 6

Robots in agriculture

Farm mechanization and precision farming

Arpana Mishra, Rashmi Priyadarshini, and Pushpa Choudhary

6.1 INTRODUCTION

This chapter provides a literature review of agriculture robotic agriculture wireless sensor networks (ARWSN), a new, innovative, and interdisciplinary area of study that is developing at the crossroads of robotics and agriculture WSNs. An ARWSN is defined in this chapter as an autonomous networked multi-robot system that uses cooperative control, learning, and adaptation to accomplish sensing goals while meeting and sustaining communication performance constraints. While work in robots and WSN has been done to a large extent, there is a wealth of untapped potential and uncharted territory at the crossroads of these two disciplines. To illustrate the benefits of controlled mobility in packet routing, consider a scenario where a pair of robotic routers establish a temporary communication channel between a transmitter and a receiver [1, 2]. Findings from this chapter indicate that while there are many publications in the robotics and WSN literatures that are pertinent to this emerging area of study, there are only a small number of articles that can be considered directly ARWSN-related. In order to make sense of everything, this chapter begins by outlining the fundamental issues and current directions of ARWSN research, including topics such as robust information flow, routing, connection, and localization.

Following the identification of issues and trends in the first stage, this chapter organizes the current state of the art in robotics and WSN research regarding ARWSN. Finally, this chapter examines gaps in the current literature and proposes future study directions. Mobile robots are widely utilized in smart factories and other forms of industrial automation [3, 4]. The workplace is a collaborative environment where different types of mobile robots work together to complete complex tasks [5]. Furthermore, WSNs in agriculture provide a virtual layer that computational systems can use to obtain data about the actual world. Additionally, mobile robots have always been an integral aspect of WSNs, and a lot of work has gone into studying how to optimize system performance with them in recent decades.

DOI: 10.1201/9781003537083-6

There are three main types of mobile robots used in WSNs, each suited to a certain set of duties:

- Gathering: when roving robots collect information from various sensor nodes;
- Transport: whereby roving robots supply the sensor nodes with resources like energy;
- A hybrid approach: whereby the mobile robots do both collecting and delivery tasks simultaneously [5, 6].

Problems can be addressed by merging concepts from various fields, according to a few major ideas. In order to facilitate the transfer of data among the mobile robot's nodes, the research presents a paradigm for a wireless network. It has several potential uses, including control and monitoring [7]. Every node has its own set of sensors and communication gear. A network of nodes is gathering information from various sensors and transmitting it to a main server. One example is a host computer linked to an online network. It is known as a WSN [8]. The main objective is to find a way to lower the computational and energy consumption of robot nodes [9].

A practical approach to the deployment of wireless sensors in agriculture, whereby a mobile robot walks along a linear backbone network with agricultural branches, deploying sensor nodes as it goes [10]. The limiting cropping load capacity of the robot forces it to constantly return to the location where all sensors are temporarily kept and reload the sensor nodes. This process causes the robot to travel the path numerous times, resulting in increased energy consumption. Reducing the traveling path and achieving the needed coverage is possible with a STPR algorithm. At the temporary starting position, all the nodes are stored, and the mobile robot proceeds to its destination while continuously loading agricultural sensors [11, 12].

6.2 AGRICULTURE SENSOR NETWORK SECURITY USING ROBOTS (ASNSR)

An innovative, open, and adaptable architecture that enhances the security of static sensor networks through the utilization of robot–sensor network interaction is introduced in this article as sensor network security utilizing robots (SNSR) [13]. Without going through any middlemen, the robot in SNSR authenticates sensor nodes and uses radio-based reallocation to build routes and conduct centralized topology calculation [14]. It then communicates directly with nodes, sending them configurations or receiving status and abnormality reports. In order to adapt to changes, respond to attacks, or detect and repair problems, SNSR operation is divided into stages placed in a feedback iterative structure [15]. This allows for the execution of stages to be repeated. High levels of security and adaptability are provided by SNSR by utilizing the capabilities of the robot, all without the need for sophisticated processes. This article introduces SNSR, discusses its defenses against typical attacks, and provides experimental evidence of its effectiveness [16].

To fulfil the most fundamental needs for coverage and connectivity, a few sensors are needed. When it has enough sensors, the mobile robot returns to the most recent branch after arriving at the far intersection and deploying them along the branches according to the algorithm rules. When it does not work, the robot turns around and puts sensors along its reverse route [17]. Once the robot has deployed all of the branches and backbone, it will reload its sensors and begin the procedure all over again. A large body of research establishes that algorithms provide a viable alternative to more conventional approaches. The technique successfully decreases the movement distance at the randomly formed network, according to the simulation findings [18].

6.3 WATER DISTRIBUTION NETWORK

In certain areas, like Texas's Rio Grande Basin, rehabilitation of irrigation districts is already under process. To manage gate automation, objectively assess irrigation water needs, usages, and losses, and to prioritize and analyze different rehabilitation solutions, water distribution network models are necessary, as discussed in this chapter. Commercially accessible software packages have limited functionality in certain applications because of their high cost and operating complexity, as discussed in this chapter. The goal of this research is to create a model that can represent the distribution network water flow profile in irrigation systems [19, 20]. Making the modeling process easy, quick, dependable, and precise was the primary purpose of building the modeling tool. With the assumption of steady, gradually changing flow, the modeling tool for branching canal networks was built based on methodological investigation. Using a single channel with a root mean squared error of 1% compared to the benchmark calculation and a branching network with relative errors ranging from 5% to 12% compared to check-point measurements along the network, the tool's flow profile computation was confirmed. In order to modernize irrigation systems, it is necessary to accurately quantify water for planning, analysis, and development purposes [21, 22].

6.4 IRRIGATION DISTRIBUTION NETWORKS

Agricultural water supply is greatly facilitated by irrigation distribution networks. Farms receive water from irrigation districts via pipelines and channels. The irrigation districts are currently undergoing rehabilitation efforts. Management of gate automation, scientific quantification of irrigation water needs, usages, and losses, and prioritization and analysis of different rehabilitation solutions all necessitate quantitative evaluation methods. Water resource planning and management software and computer models have been the subject of extensive study over the last 30 years. You can automate the gates of irrigation channels and model their flows with models and tools that are either commercially or academically available. One such example is SOBEK, developed by Delft Hydraulics in the Netherlands. It is a 1D/2D modeling program used for river, rural, and urban water management, design, planning, and policy making. Software packages like SOBEK and cavalcade are pricey and designed for general usage. On the other hand, HEC-RAS and canal Man are free to download but can be difficult to customize for individual uses. Irrigation system models have been tested. Many authors simulated models for both individual furrows and entire fields [23, 24]. Using model simulations, we assessed the significance of each spatially-varying model input to irrigation performance.

6.5 FOUR FURROW IRRIGATION MODELS

The Ross, Walker, Strew off, and Elliott furrow models are irrigation distribution networks. Under three field circumstances in southeast Australia, these models are used to forecast advance and recession timings, computing time per simulation run, and volume balance error. A method has been devised to calibrate models of on-demand irrigation networks. "Using this method, we compared three objective functions: two popular ones, and one new one. An on-demand irrigation network in Albacete, Spain, was subjected to this approach, with flow and pressure recorded at the hydrant level [25].

A hydraulic simulation model was designed for the purpose of simulating steady and turbulent flows in irrigation canal networks, according to some simulation models. To discretize the Saint-Tenant equations, the model uses the implicit four-point Pressman scheme. The following equations are then solved using the sparse matrix solution technique. This model can be used to model the flow in both straight lines and branches [26].

6.6 LOOPED CANAL NETWORKS

For most looping canal networks, the Saint-Venant equations provide a description of the unstable gradually varying flow (LCNUSGVF). These equations contain many boundary conditions and are partial differential equations that are simultaneous [27]. Putting an unstable canal model into practice calls for substantial human and material resource investments. A single ordinary differential equation explains steady gradually varied flow (SGVF), a specific instance of LCNUSGFV, which is considerably easier to construct than the Saint-Venant equations. For many flow computing and analysis issues, the SGVF description works well, and the LCNUSGVF could be reduced to cascaded SGVFs. In order to predict the water flow profile in irrigation distribution networks in the Rio Grande Basin in Texas and other comparable places, the study aimed to construct a modeling tool based on the description of SGVF [28]. A streamlined, quick, dependable, and precise modeling procedure is what you can expect from the newly created modeling tool.

In the field of study, the properties of irrigated farmland vary between regions. Specifically, this research will center on irrigated regions that exhibit the following traits:

- The waterways are shallow and have small hydraulic gradients.

In other words, the channel bottom slope is small and the water flows mildly from upstream to downstream with gravity and sufficient head pressure:

- The distribution networks are dendritically, i.e., the routes of the networks are branched but not looped.
- The networks are open channels in the Lothis Rio Grande Valley in Texas (Figure 6.1), the project area, the elevations range from sea

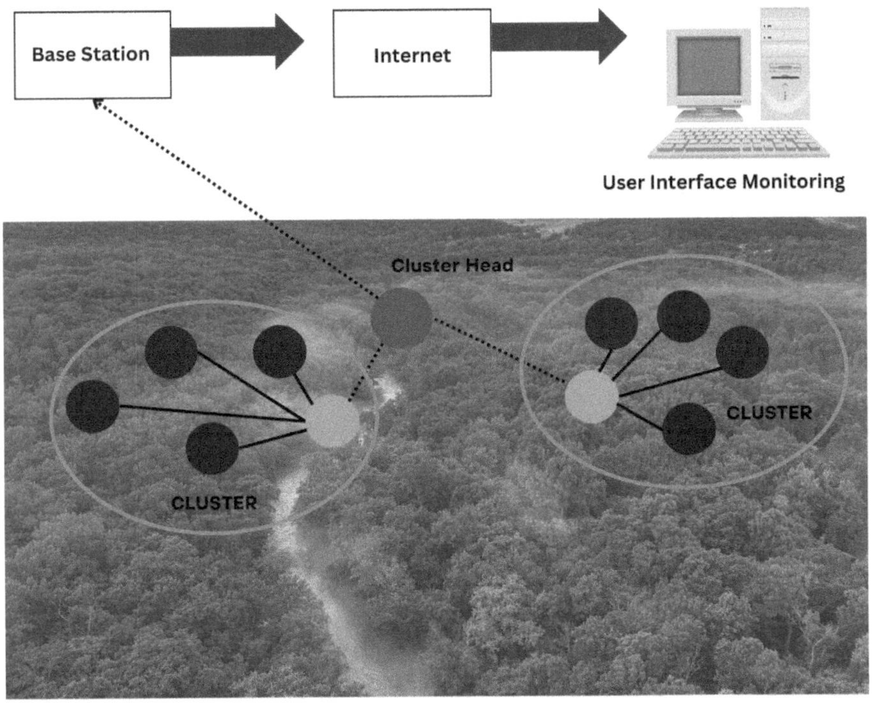

Figure 6.1 Agriculture robotic wireless sensor networks (ARWSN).

level in the east to about 200 m in the north but are mainly less than 100 m.
- Much of the area is nearly level.
- Drainage ways are shallow and have low gradients.

The canals and pipelines in the distribution networks have small three computing methods. In the open-channels (canals) of irrigation networks, water flows are typically categorized as follows:

(1) Steady uniform flow (SUF)
(2) Steady gradually varied flow (SGVF)
(3) Unsteady gradually varied flow (USGVF)

The SGVF can be calculated and studied by applying an ordinary differential equation to the observation of mass and energy conservation. The Saint-Venant equations, which take into account the conservation of mass and momentum, can also be used to calculate and analyze the USGVF. The mathematical derivation of the SGVF as a subset of the USGVF is straightforward [29, 30]. Because they are partial differential, the Saint-Venant equations make computation implementation considerably more challenging. The SGVF has proven to be an excellent and practical solution to numerous issues related to flow calculation and analysis. Applying various beginning and boundary hydraulic conditions to the basic equation allows for the cascade of solutions along a canal channel and distribution network structure. In most irrigation systems, the flow conditions are not uniform [31, 32].

In regions with short irrigation canals and low hydraulic gradients, like Texas's Rio Grande Basin, SGVF is the dominant flow type, however USGVF can occasionally occur due to transient processes occurring near gate structures. Hence, the method required to construct the modeling tool is the computation of SGVF profiles in irrigation distribution networks [33].

6.7 SGVF FLOW PROFILE COMPUTATION

The computation of the SGVF profiles basically solves the governing ordinary differential equation. The main objective of the computation was to determine the shape of the flow profile. Broadly, three methods of the computation are classified as

- The graphical integration method
- The direct-integration method
- The step method

Although the graphical-integration technique is simple and easy to understand, it could be somewhat time-consuming when applied to real-world challenges [12]. Reason being that not all channel cross sections allow for an explicit expression of the SGVF differential theory [34].

Due to the impossibility of performing a direct and accurate integration of the equation, this method has only been used to solve the equation in some rare instances or to incorporate assumptions that allow mathematical integration to be executed.

The basic idea behind a step approach is to split a channel into shorter reaches and then iteratively carry out the computation along those reaches [17]. There is a wide range of step-by-step approaches. No one has been determined to be the greatest in all applications, even though some seem to be better than others in some aspects. In order to calculate flow profiles for irrigation channels, this paper lays out a step-by-step process [35]. This approach partitions a channel into smaller ones. Iterative procedures can fail if the reach lengths are too large, while computing cost will grow if they are too little. When dealing with subcritical flow, the calculation begins at the downstream end of the channel; when dealing with supercritical flow, it begins at the upstream end. This is done by applying the Bernoulli equation.

6.8 PRECISION FARMING

Precision farming makes use of agricultural parameters that can be tracked through the Internet of Things (IoT). For improved environmental protection and long-term agricultural production sustainability, the IoT delivers critical data pertaining to crop, soil, water, and air conditions in real-time. It is possible to quickly increase the efficiency of irrigation by converting it into a smart irrigation system [22]. In order to quickly and accurately gather data on vital agricultural elements such as pest attack, soil quality, and water availability, equipment like unmanned aerial vehicles (UAVs) are utilized. In order to identify pest infestations in farms, UAVs equipped with spectrum analysis technology can take high-resolution pictures and store them in databases on cloud servers. This allows for further analysis even from faraway places [14]. A smart greenhouse is a new kind of greenhouse that can grow plants with little to no human involvement. In order to improve production, it entails constantly monitoring soil moisture, temperature, humidity, and light using sensors. Decisions can be made to conduct additional corrective actions to avert agricultural harm with the help of these automated sensors [15].

Applying the aforementioned algorithm to various beginning and boundary hydraulic circumstances, one can compute SGVF flow profile in a distribution network or canal channel by cascading the solutions step-by-step along the channel or distribution network structure. The regions under consideration are mostly served by irrigation distribution networks that branch out. In most cases, laterals, second-level laterals, and even third-level laterals run the length of a primary canal in such networks. Initializing the discharge and water depth at one end of the main canal is the first step in the flow profile computation over a branching network. Then, when the calculation reaches a lateral, it must proceed by setting the initial values for the water depth and discharge at the end of the lateral [22]. The computation must proceed by initializing the discharge and water depth at one end of the second-level lateral when it progresses to a second-level lateral. When you reach the farm turnouts, you will need to recursively return to the main canal to continue the computations. As soon as it reaches the main canal's opposite end, the computation will terminate [24].

Subcritical SGVF profile computation across a branching irrigation network is illustrated in Figure 6.2. Branching irrigation networks in any shape can be handled by this approach, provided that the first-level laterals are present. Any arbitrary branching network with second-level, third-level, or *n*-level laterals can be simply expanded to using this approach. The most common type of structure used to regulate the flow of water in irrigation channels is the gate. Typically, gate configurations can experience one of four distinct flow regimes. A standard equation characterizing the flow through the gate construction is available for each of the four regimes [26].

While theoretically water flow can switch between regimes, in reality many channel constrictions and structures, such flumes, only allow one flow regime to operate at a time.

Standard deviations of residuals are a good measure of the calibration's efficacy.

The modeling tool and the program that was built are used in the modeling tool prototyping technique. Three modules have been designed to streamline, accelerate, and improve the accuracy of the modeling process:

(1) SGVF computation for a single canal channel
(2) SGVF computation for branching canal networks
(3) Flow computation through control sections

To calculate the flow through flumes and gates, use the third module. The calibration and saving of the discharge and depth relationships for use in the model. For the modeling tool's prototype, the C++ programming language was selected [29]. The applications created and built according to object-oriented programming (OOP) standards.

The data from a single irrigation canal channel are used as a baseline in model validation, together with water flow profiles from a branching network irrigation scheme. The computation of this study was validated by comparing the results with Chow's computation [27, 23]. Located in an irrigation sector in Jamaica, this system is a real-life network of irrigation

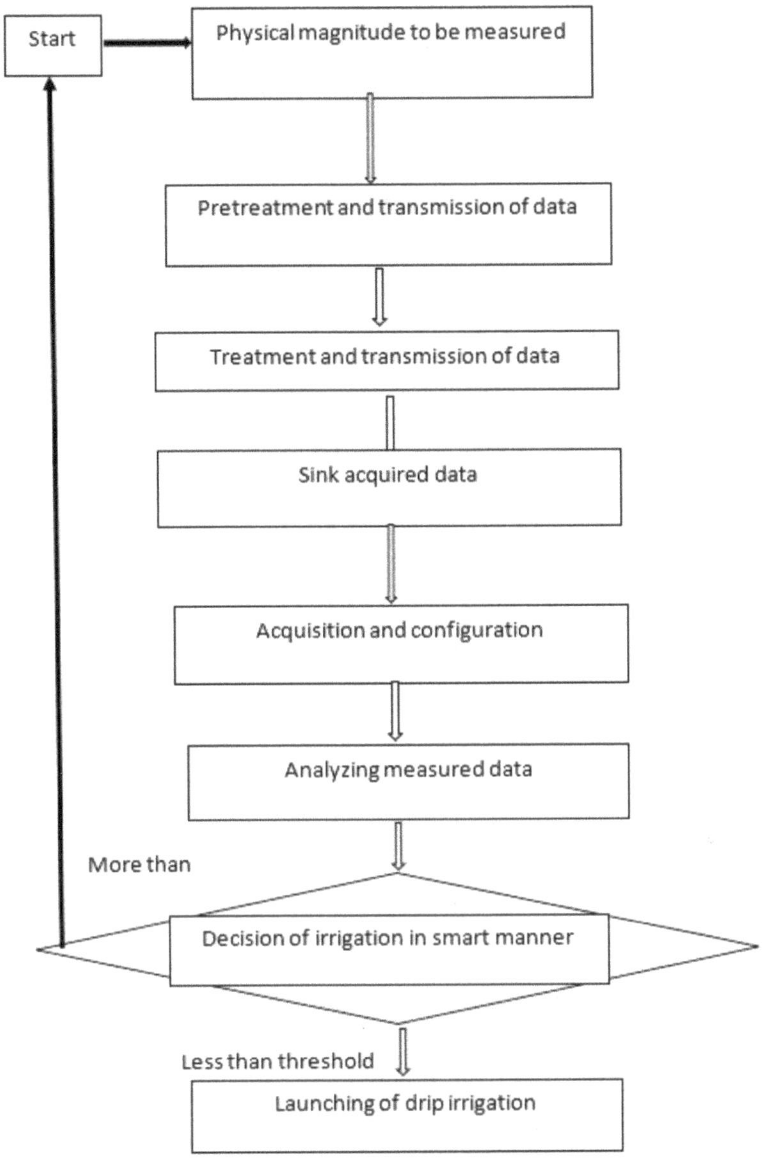

Figure 6.2 Subcritical SGVF profile computation flow over first-level lateral branch irrigation networks.

canals that stretches for around 1700 meters and encompasses an area with similarities to Texas's Rio Grande Valley. Information gathered from flow measurements and field surveys [22, 28].

6.9 IRRIGATION DISTRIBUTION NETWORKS

In order to predict the water flow profile in irrigation distribution networks, this study has developed a tool. It is assumed in the modeling tool that the branching canal networks have SGVF flow. Before beginning the calculation, the tool sets the initial values for the discharge and water depth at the main canal and lateral ends [18, 13]. It manages first-level lateral-branching networks in arbitrarily structured topologies. This approach may be modified to handle arbitrary branching networks with laterals at any level from 2 to *n*. A single irrigation canal channel and an irrigation branching canal network were used to assess the modeling tool's performance in water flow profiling. The modeling tool's results were 1% off from the benchmark for a single channel flow and 5% to 12% off from check point measurements for a network of branching canals. In comparison to commercially available models and software packages, the modeling process becomes simple, fast, reliable, and accurate with the least amount of configurations and expense thanks to the modeling tool's strong capability in handling the modeling task in different conditions, as demonstrated by its implementation and results. The results of this study will be useful for irrigation districts in the Rio Grande Valley of Texas and other comparable places when it comes to water quantification for planning, analysis, and development of contemporary irrigation systems. The field survey and flow measurement were conducted by Mr Milton Henry of the Jamaican National Irrigation Commission. The writers are grateful to him for his assistance [15, 17].

6.10 QUALITY INDICES IN AGRICULTURE

The decline in agricultural labor in China is making the country's once-famous tea production unsustainable and driving up prices. This study proposes a self-propelled track-type tea plucking machine that combines mechanics, hydraulics, and electronics to enhance the mechanization of tea plucking. The primary characteristic according to tea garden conditions is finite element software to develop and analyze the vibration performance of three distinct types of cutter blades. A rigid-flex mixed motion model was also developed. First, the model and plots make it clear that the blades move in a sine law. Second, as they travel, the blades undergo a symmetrical deformation from the middle to the end. Third, the deformation grows as it moves along the blade, from the center to the sides. Quality indices (integrity rate, unpicking rate, stubble unevenness) of plucked tea leaf were found to be significantly affected by both travel speed and the speed ratio of cutting and traveling, according to the results of the single factor experiment [33].

The regression equations of three indices regarding two experimental elements using the universal rotary combination experiment of quadratic regression. In addition, a software design expert optimized all three indexes thoroughly. Here are the ideal settings for the machine: the three indices forecast that the integrity rate may achieve 78.26%, the unpicking rate lowered to 0.82%, and the stubble unevenness reduced to 1.30 mm, with a travel speed of 0.41 m/s and a speed ratio of cutting and traveling of 1.06. Stubble unevenness decreased to 1.23 mm, unpicking rate to 0.87%, and integrity rate to 77.41% according to the results of the verification trial. Each of these test indexes exceeds the requirements laid out in the machine-picked tea standard, and they are all rather near to the projected values. The widest possible application of this tea plucking machine is made possible by the synthetically optimal parameter set. It is suitable for use in tea gardens with lengthwise slopes less than 18° and crosswise slopes less than 20°, as well as tea tree heights between 50 and 120 cm. As we can see, the current tea picking problem is a bottleneck in the tea business, and the tea plucking machine that this study investigates offers a viable solution. Improving the machine's automation and intelligence will require more research [32]. It is important to harvest mature tea at the right time, as discussed in this chapter, because otherwise the nutrients would be used up for growth, resulting in poor quality. In addition, the plucking period is typically shorter than seven days. Tea picking has always required a large number of human workers. Agriculture is seeing a dramatic decline in labor force participation as a result of societal progress. As a result, tea plucking machines are essential and will give a foundational boost to the tea industry in China and other regions that produce tea [23].

Research into the potential mechanization of tea plucking is a relatively recent phenomenon. The mechanical and physical features of young tea sprouts, as well as their design and kinematics analysis, have been the primary subjects of relative investigations. Japan is a major supplier of traditional tea plucking machines to China and other producing areas [35]. There are two main categories of these machines, based on their power source (electrical motor or petrol engine), and three different ways they move (one operated by a single person, one operated by two people, and a self-propelled model). All of these machines collect tea leaves from the tree's crown, but they lack the ability to regulate the length and sensitivity of the leaves to meet the standards of China's renowned tea. Notable examples of common single-type include 4CD-20 (electrical), 4CD-330 NV45H, AM-100E, ML-500, and so on. Some examples of common du-type viruses are 4CSW910, 4CSW-1000, PHV-100, V New Z-1000, and others. Due to the lack of selection, these tea plucking machines are unfit for use in the world of famous tea plucking. A handle-controlled, self-propelled tea plucking machine was recently created. It increased production to a certain degree, although it is not likely to be utilized frequently. The tea-making process in Japan is highly mechanized; for example, a popular tea plucking machine is a self-propelled riding type one, which greatly improves efficiency and automation [19].

6.11 INTEGRITY RATE IN GARDENS

The lack of selectivity in the plucking process means that this machine is only suitable for harvesting large quantities of tea, not high-quality varieties. The low cost and regional variations in agricultural practices also prevent it from finding widespread use in China and other tea-producing regions. Domestic tea planting areas have long endured the arduous task of collecting tea throughout the plucking season. Specifically for household tea gardens with a gentle slope, this study proposes a self-propelled track-type tea plucking machine to automate the process of picking high-quality tea more efficiently in China's tea gardens. This chapter details experiments performed in a tea garden to optimize parameters and achieve complete optimization, with the goal of improving the integrity rate. An effective mechanical device for high-quality tea plucking in tea gardens with gentle slope has been developed through this research. This device can significantly reduce the labor pressure currently associated with tea plucking. The tea plucking machine is based on the arc-type double-blade reciprocating cutting principle and primarily includes a track chassis with hydraulics, a frame assembly, a plucking system, a system for collecting leaves, a system for transmitting power to the pother, a system for raising the cutter, and so on [11, 17].

6.12 QUALITY PLUCK USING ROBOT

The rubber tracker and frame assembly make up the machine's main body. At four points along the escalator's hinged length, the plucking motor, collecting bags and blast pipe are fastened to the lift mechanism's sprocket chain. A flexible pipe links the blast pipe to the draught fan, which in turn guides the wind upward to the cutter, which gathers the sprouts that have been plucked. As the cutter sits atop the tea trees, the machine bobs along their branches, picking off the sprouts of tea as they grow. To transfer the selected tea leaves to the collecting bags, draught fans are utilized. After the tea sprouts are plucked, they should be placed in a specific container [27]. Adjusting the lift system's height with a motor allows for precise positioning of the cutter in relation to the tree's surface, which in turn guarantees high-quality plucking.

Agriculture wireless sensor networks (WSNs) give computational systems access to data about the physical world through a virtual layer [7]. For quite some time, mobile robots have been an integral part of WSNs, and over the past few decades, a great deal of research has focused on finding ways to improve system performance by utilizing mobile robots. The use of mobile robots in WSNs is discussed in this study along with various strategies in the field.

The literature regarding agriculture WSN can be divided into three groups based on the activities carried out by mobile robots: the first is "collection," in which mobile robots gather data from sensor nodes; the second is "delivery," in which they provide sensor nodes with resources like energy; and the third is "combination," in which they do both tasks at once. Here we detail the current solutions in relation to the suggested taxonomy. Detailed descriptions and analysis are provided to assist readers understand the flow within each category. The chapter also outlines the link among potential solutions. The chapter also compares and contrasts the available options across different categories in terms of mathematical formulation, application, etc., showing how they differ and are similar. Lastly, this chapter has been successful in identifying a number of unanswered or under-researched questions.

6.13 INTERDISCIPLINARY FIELD OF WIRELESS SENSOR NETWORKS (WSN)

The primary impetus for the interdisciplinary field of wireless sensor networks (WSN) in agriculture came from military uses. New uses are popping up all the time, thanks to advancements in technology and the falling production costs of sensor nodes. Some examples include habitat monitoring and intrusion detection. Wireless sensor networks (WSNs) are a potentially useful tool for gathering data from a variety of physical sources, which in turn might aid in raising environmental consciousness. The efficient collection of distributed sensory input is a major concern in WSNs [3, 4, 5].

Traditional WSNs in agriculture include a network of wireless sensor nodes that gather data about their immediate surroundings and transmit it to a central data center, sometimes called a sink or base station (BS), where it may be processed and analyzed. Typical WSNs have static sensor nodes and sinks that communicate with one another to transfer sensory data from one node to the other. Because they are responsible for both transmitting and relaying sensory input, the nodes closest to the sinks tend to be overloaded, leading to the energy hole problem. When this happens, the nodes in question often experience rapid battery drain, cutting them off from the rest of the network. Designing energy-efficient communication protocols is just one example of the many efforts performed to increase network efficiency in this area. Regardless of what happens in this chapter, the energy hole problem will persist until the sinks and sensor nodes are dynamic.

An efficient way to alleviate the energy hole issue and then improve the network lifetime is by introducing mobile nodes into WSNs [4]. They can assist static sensor nodes in relaying sensory input by moving closer to them. They can be fastened to vehicles or robots and sent to sensor nodes to gather data at short range, cutting down on power consumption at such locations. Due to the restricted energy supply at sensor nodes, the network lifetime is still constrained, even when it is lengthened. Scientists have pondered the question of how to construct eternal networks while avoiding this limitation. Allowing sensor nodes to absorb energy from their surroundings is one possibility. In any case, the environment plays a significant role in this method. One novel method has emerged as a result of wireless packet transfer technology. Placing a large enough battery on a robot would allow it to function as a portable charger, allowing users to wirelessly top off their batteries. Since a robot's power consumption accounts for a disproportionately large portion of overall power consumption, it may be more efficient to use a single robot to collect and charge energy rather than to request two separate robots. As shown in Figure 6.1, this insight gives rise to a novel challenge: the simultaneous gathering of data and charging of energy by mobile robots.

6.14 MOBILE ENTITY

The term "mobile entity" (ME) is used to describe a mobile sink in the data collection approaches. Data collecting and recharge are only illustrative of the two concepts, respectively, of collection and transmission. Among the many possible uses for the technologies are the following: dynamic key generation (collection and delivery), software update (delivery), detection of malicious nodes (collection and delivery), etc. [14, 15].

The role of MEs in WSNs has been the subject of voluminous study. In WSNs, there are existing polls that cover specific ME-related subjects, so covering, transferring, and routing are only a few of the many topics addressed in [7]'s comprehensive study of data collecting with MEs. A number of mobility-aware MAC protocols can be compared in the work of AWSN.

Various AWSN algorithms uses the taxonomy and the properties of sensor nodes utilized as MEs in device applications. Additionally, this AWSN explores the common aims of existing strategies for ME usage in WSNs, as well as the various limits that these methods consider. By referencing these discussions in the sections that follow, we can distinguish between the different approaches.

6.15 COLLECTION TASK

The most important use case for collection activities is using MEs to acquire data from sensor nodes. The three potential components of such a system are a stationary base station (BS), a mobile edge server (ME), and a group of fixed sensor nodes. The freedom of movement of the MEs in the field have the ability to move under partial control. They can move completely under control in a single hop technique. Single hop is one way that sensor nodes can communicate with MEs about the data that they have detected [17].

6.16 DELIVERY AND COMBINATION TASKS

Several articles that are pertinent to the delivery task and the combination task. Using MEs to wirelessly charge sensor nodes is the key use in the delivery task category. Additional uses exist, such as content distribution. For instance, MEs communicate with sensor nodes by relaying commands and software updates". Energy delivery methods are presented in content delivery methods [11].

6.17 CONCLUSION

Adding mobility as an extra component improves the efficacy of WSN in several ways. A lot of effort has gone into developing networks that consider WSN properties, and there have been a number of survey publications on WSNs with MEs. The primary focus of these surveys is data collection and relay; these duties are included in the suggested taxonomy of MEs. Delivery and combination make up the remaining two sections of the taxonomy. Here we outline the format for the literature reviews.

Pest monitoring in smart farming used on agricultural areas, equipment that generate ultrasonic sound waves can deter a variety of pests. Powerful ultrasonic sound waves are generated by an electronic insect repellent. Pests such as mosquitoes and rats are scared away by these waves. Consequently, these tools are a great substitute for conventional pesticides, which have many negative effects on helpful creatures. Another factor that can lead to increased use of ultrasonic equipment is accidental contact to chemical insect repellents. There is no harm to humans or the environment caused by electronic insect repellants, and they are inexpensive. For the most part, these devices work by appealing to the pests' auditory senses.

REFERENCES

1. Aahin, E., Winfield, A.: Special issue on swarm robotics. *Swarm Intell.* 2(2–4), 69–72, 2008.

2. Nagel, R., Eichler, S., Eberspacher, J.: Intelligent wireless communication for future autonomous and cognitive automobiles. In: *Proceedings of the IEEE Intelligent Vehicles Symposium (IV),* pp. 716–721, 2007.
3. Mishra, A., Priyadarshini, R., Mehra R.M.: Optimized energy clustering routing Protocol-WL. *J Inform Optim Sci. (WoS & ABDC)*, **44**(1), 67–80, 2023. https://doi.org/10.47974/JIOS-1296
4. Kovács, T., Pásztor, A., Istenes, Z.: Connectivity in a wireless network of mobile robots doing a searching and collecting task. In: *Proceedings of the IEEE International Symposium on Applied Computational Intelligence and Informatics (SACI),* pp. 716–721, 2009.
5. Mishra, A., Priyadarshini, R., Mehra R.M.: 'IOT based smart sensing health monitoring', *National Conference on Recent Trends in Academic Research (RTAR-2021)* organized by Council of Industrial Innovation and Research, (CIIR), Delhi, 30–31 August, 2021.
6. Du, Z., Yang, W., Chen, Y., Sun, X., Wang, X., Xu, C.: Design of a robot cloud center. In *Proceedings of the International Symposium on Autonomous Decentralized Systems (ISADS),* pp. 269–275, 2011.
7. Baeg, S.-H., Park, J.-H., Koh, J., Park, K.-W., Baeg, M.-H.: Building a smart home environment for service robots based on RFID and sensor networks. In: *Proceedings of the IEEE International Conference on Control, Automation and Systems (ICCAS),* pp. 1078–1082, 2007.
8. Mishra, A., Priyadarshini, R., Mehra R.M.: 'Task decomposing optimization in in wireless sensor network', *1st International conference on SCI/ TECH and Engineering (ICSTE-2023)*, organized by National Institute of Technology Manipur, 17–18 Feb,2023. *Lecture Notes in Electrical Engineering*, pp. 453–462, 2023.
9. Stilthis chapterll, D.J., Bay, J.S.: Toward the development of a material transport system using swarms of ant-like robots. In: *Proceedings of the IEEE International Conference on Robotics and Automation (ICRA),* 1993.
10. Baber, J., Kolodko, J., Noel, T., Parent, M., Vlacic, L.: Cooperative autonomous driving: Intelligent vehicles sharing city roads. *IEEE Robot. Autom. Mag.* **12**(1), 44–49, 2005.
11. Liu, H., Darabi, H., Banerjee, P., Liu, J.: Survey of wireless indoor positioning techniques and systems. *IEEE Trans. Syst. Man Cybern. Part C: Appl. Rev.* **37**(6), 1067–1080, 2007.
12. Murphy, R.R., Kravitz, J., Samuel, L.S., Shoureshi, R.: Mobile robots in mine rescue and recovery. *IEEE Robot. Autom. Mag.* **16**(2), 91–103, 2009.
13. Ghosh, P., Ren, H., Banirazi, R., Krishnamachari, B., Jonckheere, E.: Empirical evaluation of the heat-diffusion collection protocol for agriculture wireless sensor networks. *Comput. Netw.* **127**, 217–232, 2017.
14. Wang, S., Gasparri, A., Krishnamachari, B.: Robotic message ferrying for wireless networks using coarse-grained backpressure control. *IEEE Trans. Mob. Comput.* **99**, 1–1, 2016.
15. Mishra, A., Priyadarshini, R., Mehra R.M.: 'Qualitative analysis of intraclass and inter class clustering routing and clusterization in wireless sensor network', Evergreen, Transdisciplinary Research and Education Centre for Green Technologies, Kyushu University, 8(2), 358–373, 2021. https://doi.org/10.5109/44807182

16. Mishra, A., Hassan, H.M., Tiwari, U., Priyadarshini, R.: Subjective survey on probabilistic and non-probabilistic clusterization in wireless sensor network, *Wireless Pers Commun.* **132**(3), 1703–1729, 2023. https://doi.org/10.1007/s11277-023-10629-4
17. Mishra, A., Priyadarshini, R., Mehra R.M: 'A Qualitative analysis on Application based different routings in wireless sensor network', *7th International conference on trends in electronics and informatics (ICOEI-2023)*, Tiruvanevli, Manipur, 11-13 April, 2023. https://doi.org/10.1109/ICOEI56765.2023.10125695
18. Rana, S., Jain S., Kanji, R.: 'Design of multi chain power efficient gathering for sensor information system using sink mobility', *Sixth International Conference on Image Information Processing (ICIIP)*, Shimla, India, 2021, 81–86. https://doi.org/10.1109/ICIIP53038.2021.9702634
19. Randriatsiferana, R.S.A., Lorion, R., Alicalapa, F., Harivelo, F.: 'Energy-efficient clustering algorithm based on energy variance for Wireless Sensor Networks', *International Conference on Smart Communications in Network Technologies (SaCoNeT)*, Paris, France, 2013, 1–5. https://doi.org/10.1109/SaCoNeT.2013.6654571.
20. Reshme, A.: 'Delay and energy efficient sata collection in wireless sensor networks', *International Conference on Recent Advances in Energy-efficient Computing and Communication (ICRAECC)*, Nagercoil, India, 2019, 1–6. https://doi.org/10.1109/ICRAECC43874.2019.8995170
21. Rostami, A.S., Badkoobe, M., Mohanna, F.: Survey on clustering in heterogeneous and homogeneous wireless sensor networks, *J Super Computer.* **74**, 277–323, 2018. https://doi.org/10.1007/s11227-017-2128-1
22. Sachan, S., Vats, M., Mishra, A., Choudhary, S.: Comparative analysis of clustering algorithm for wireless sensor networks, *Int Comput Eng.,* **1125**, 63–71, 2020. https://doi.org/10.1007/978-981-15-2780-7_9
23. Shen, J., Wang, A. Wang, C., Hung P.C.K., Lai, C.F.: An Efficient Centroid-Based Routing Protocol for Energy Management in WSN-Assisted IoT, *IEEE Access*, 5, 18469–18479, 2017. https://doi.org/10.1109/ACCESS.2017.2749606.
24. Shi, D., Hui, G., Zhang, X.: An energy-optimization clustering routing protocol based on dynamic hierarchical clustering in 3D WSNs, *IEEE Access*, 7, 80159–80173, 2019. https://doi.org/10.1109/ACCESS.2019.2923882.
25. Singh, K.: 'WSN LEACH based protocols: A structural analysis', *5th International Conference and Workshop on Computing and Communication (IEMCON)*, Vancouver, BC, Canada, 2015, 1–7. https://doi.org/10.1109/IEMCON.2015.7344478
26. Sinha, A., Anantho, C.: Dynamic power management in wireless sensor networks, *IEEE Des Test of Comput.* **18**(2), 62–74, 2001.
27. Sivakumar M. and Divakaran, J.: 'Load balancing based energy efficinet routing approach for heterogeneous wireless sensor networks', *Online International Conference on Green Engineering and Technologies (IC-GET), Coimbatore, India*, 2015, 1–12, https://doi.org/10.1109/GET.2015.7453849
28. Sun, W., Song, Y., Chen, M.: 'A load-balanced and energy-aware routing metric for wireless multimedia sensor networks', *IET 3rd International Conference on Wireless, Mobile and Multimedia Networks (ICWMNN 2010), Beijing*, 2010, 21–24. https://doi.org/10.1049/cp.2010.0608

29. Sun, Z., Liao, G., Zeng, C., Lan, L., Zhao, G.: GLBR: A novel global load balancing routing scheme based on intelligent computing in partially disconnected wireless sensor networks, *Int J Dist Sensor Net.* **18**(4), 2022. https://doi.org/10.1177/15501329221090458
30. Tang, M., Lin, X., Palesi, M.: Routing pressure: A channel-related and traffic-aware metric of routing algorithm, *IEEE Transact Para Dist Syst,* **26**(3), 891–901, 2013.
31. Tarachand, A., Kumar, V., Raj, A., Kumar, A., Jana, P.K.: 'An Energy efficient load balancing algorithm for cluster-based wireless sensor networks', *Annual IEEE India Conference (INDICON)*, Kochi, India, 2012, 1250–1254. https://doi.org/10.1109/INDCON.2012.6420810
32. Tellioglu, I., Mantar, H.A.: 'A proportional load balancing for wireless sensor networks', *Third International Conference on Sensor Technologies and Applications*, Athens, Greece, 2009, 514–519. https://doi.org/10.1109/SENSORCOMM.2009.86
33. Wang, A., Heinzelman, W.R., Chandrakasan, A.P.: 'Energy-scalable protocols for battery-operated microsensor networks', IEEE Workshop on Signal Processing Systems. *SiPS 99. Design and Implementation (Cat. No.99TH8461)*, Taipei, Taiwan, 1999, 483–492. https://doi.org/10.1109/SIPS.1999.822354
34. Wang, H., Peng, D., Wang, W., Sharif, H., Chen, H.H.: Cross-layer routing optimization in multirate wireless sensor networks for distributed source coding-based applications, *IEEE Trans Wireless Commun.* 7(10), 3999–4009, 2008.
35. Wang, H., Tu, P., Wang, P., Yang, J.:'A redundant and energy-efficient clusterhead selection protocol for wireless sensor network', *Second International Conference on Communication Software and Networks*, Singapore, 2010, 554–558. https://doi.org/10.1109/ICCSN.2010.46

Chapter 7

Smart autonomous transport

Land and water

Muhammad Abdullahi Musa and T. Jayachitra

7.1 INTRODUCTION

Recent advances in technology have revolutionized the way we travel and move goods. Rapid technological advancement is currently occurring in the automotive industry, with connected networks, computers, software, and sensors being utilized more frequently to make cars safer for both drivers and passengers. Smart autonomous transport is a driverless form of transportation made possible by autonomous vehicles and autonomous vessels. Autonomous vehicles were described as, "computers on wheels, a revolution akin to the move from the horse and buggy to the automobile" [1]. In the same vein, the maritime industry is witnessing increasing efforts toward placing unmanned ships on the sea and inland waterways. Smart autonomous transport systems offer efficient and sustainable ways for land and water transport. This chapter delves into the realm of smart autonomous transport in land and water with a particular focus on its application and future trends.

Smart autonomous transport systems were made possible owing to the extensive researches and developments on the frontiers of smart wireless sensors, big data analytics, energy harvesting, nanotechnology, high-performance computing, imaging, and visualization. These systems make use of algorithms to automate the movement of vehicles and vessels without human intervention. They enable the vehicles to navigate safely, interpret data precisely, and make decisions in real time efficiently.

The introduction of autonomous vehicles has given rise to two primary models: one is a completely autonomous vehicle that is simpler, more fluid, effortless, and auto-adaptive, and the other is a new kind of management that ensures complete safety and security (cyber-X, physical safety, etc.) in our everyday environment under all conditions [2]. There has been a lot of interest in the development of autonomous vehicles, providing intriguing applications in shipping, marine transportation, and recreational boats, and they may be used for both land and ocean transportation. Autonomous transportation has several advantages. Because they lessen human mistake

DOI: 10.1201/9781003537083-7

[3] and the hazards associated with human weariness or distraction, these technologies have the potential to improve safety [4].

This chapter will address the many aspects of intelligent autonomous land and maritime transportation. We will look at the current state of technology and discuss its possibilities as well as the challenges that come with using it. We will examine real-world examples, such self-driving cars and self-navigating boats, in order to understand the potential impacts of these advancements on the transportation industry and society throughout.

7.2 AUTONOMOUS VEHICLES ON LAND

The advent of autonomous vehicles on land marks a paradigm shift in the way we conceive, design, and interact with transportation systems. This revolutionary technology broadly encompasses a spectrum of vehicles, from self-driving cars and trucks to, shuttles, buses, and autonomous trains [5]. These vehicles have the capability of identifying their surroundings and navigating entirely or partially without human input. The evolution of autonomous land vehicles is propelled by merging cutting-edge technologies with sophisticated sensors, machine learning, and AI [6]. Autonomous vehicles are regarded as subcomponents of autonomous intelligent vehicles, and have greater possibility of actualization in the automobile's world. They are known as "vehicles of tomorrow" (also known as the Internet of Vehicles, or IoV) [7].

Autopilot systems for airplanes were the inspiration behind the concept of driverless cars years ago [8]. The 1960s saw the development of a "self-driving" Citroen DS vehicle in Berkshire at the Transport and Road Research Laboratory. The researchers designed the vehicle to follow magnetic cables implanted in the road. There was initial successful displays in the 1980s, followed by significant advancements for self-driving cars. Autonomous ground vehicles (AGVs) have experienced upheavals, progressing through five levels of autonomy based on technological gains. The automation of vehicles is growing as a result of enhanced sensor technologies, artificial intelligence (AI) methods, and intelligent control systems [9].

The advancement of the Internet of Things, programmable logic controllers (PLC), and computer advances are making autonomous cars a reality [7]. The development of autonomous ground vehicles was a vision that begun since the first quarter of the 20th century. A major breakthrough came in 1948 when modern-day cruise control was developed. Recent developments estimate that by 2030, about half of all new cars will be driverless [9] Prototype vehicles for autonomous vehicles have already been tested for millions of miles [10].

Applications for the first three autonomy levels including hands-on support with driver assistance, hands-off with an autopilot needing continuous attention, and eyes-off still requiring human intervention in emergency situations have shown success. Advanced driver assistance systems (ADAS) and level 2 systems like Tesla's Autopilot are examples of semi-autonomous modes that call for further driver aid systems and driver attentiveness. Although they gather an immense amount of data, these systems exclusively mark the beginning of fully autonomous vehicles. Successful attempts have been made in the last two levels, where no human intervention is required [9].

7.3 AUTONOMY LEVELS

There are five stages of automation, ranging from completely automated, self-driving cars that require passengers to give no attention to the road to driver aid systems that can assist with navigation [1]. The Federal Automated Vehicle Policy of the National Highway Traffic Safety Administration (NHTSA) established these [11].

7.3.1 Level 0 (momentary driver assistance)

The driver must guide the vehicle, brake, and speed up. The vehicle has features that only assist in the operation and do not take hold of the vehicle's maneuver. Examples of the vehicle features at this level consist of automatic emergency brakes, forward collision warning, and lane changing alert. While the driver is fully engaged with close attention, the system provides temporary driving help, such as caution and warnings or emergency safety measures.

7.3.2 Level 1 (driver assistance)

While the driver is completely focused and engaged, the system continuously assists with control of the vehicle, braking, and acceleration. The vehicle must be driven by the driver. The system can operate in two ways when it is engaged: steering or acceleration/braking. Lane keeping assistance and adaptive cruise control are two examples of automotive technologies in this level.

7.3.3 Level 2 (additional assistance)

The driver stays completely engaged with their focus as the technology continuously assists with steering, braking, and acceleration. Operating the car is the driver's responsibility. When activated, the system is capable of both directing and stopping and propulsion.

7.3.4 Level 3 (conditional automation)

While the driver is still available to take over, the system actively handles driving tasks. When the system is active, it takes care of every part of driving, and you, the driver, can take over at any time. When the system malfunctions and notifies the driver, the driver needs to be accessible to continue with every aspect of the driving task.

7.3.5 Level 4 (high automation)

Within restricted-service regions, the system is entirely in charge of operating the vehicles; occupants merely function as passengers and are not required to be involved. When the system is activated, you, the passenger, are relieved of all driving responsibilities as the system takes care of everything. The car can only be operated by the system in certain service locations; it cannot operate anywhere. It can be operated without a human driver.

7.3.6 Level 5 (full automation)

Riders merely function as passengers and have no requirement to be involved; the entire vehicle operation is managed by the system. When the system is activated, you, the passenger, are relieved of all driving responsibilities as the system takes care of every circumstance. Under any condition and on all types of roads, the system is capable of operating the car globally. It can be operated without a human driver.

An autonomous vehicle uses a variety of technologies and sensors, including adaptive control, anti-lock braking systems, GPS navigation, active navigation, lasers, and radar, to navigate in "autopilot" mode. Autonomous vehicles are automobiles or trucks that do not require human intervention for safe and secure operation. Often referred to as "driverless," these vehicles combine software and an array of sensors to navigate and steer the vehicle [4].

As vehicles adopt higher autonomy levels, passenger and pedestrian safety becomes paramount. Many cars today come equipped with driver assistance technologies, which help prevent accidents and save lives on the road. Certain driver assistance technologies are meant to alert you when there is a chance of an approaching collision, while others are meant to intervene to prevent a collision. The continuous advancement of automotive technology aims to deliver even more safety advantages. Examples of this include automated driving systems and driver support technologies [11].

The potential increase in road safety is one of the main drivers underlying the emergence of autonomous vehicles. Traffic accidents in 2021 have claimed the lives of nearly 1.19 million, translating to approximately 15 deaths in every 100,000 population of people [12]. Approximately 94%

of traffic accidents are the result of driver errors [1]. They are mitigated through the precision and consistency of autonomous systems. These vehicles adhere rigorously to traffic rules, maintain optimal speed limits, and respond instantaneously to unexpected events, thereby minimizing the likelihood of accidents.

Moreover, the interconnected nature of autonomous vehicles facilitates communication between them, generating the possibility of synchronized traffic control. This can improve vehicle movement generally, greatly lessen traffic congestion, and help create more effective transportation networks [10].

Human mistake is the root cause of the majority of accidents. Vehicle safety will improve by eliminating human judgment and split-second decision-making. Various aspects would make AVs safer, or at least perform better than humans. The multiple sensors around an AV will be more aware of its surroundings than a single human driver. The capacity to notice changes in the environment faster than a person would help keep people safe. However, there is a flip side. What will an autonomous vehicle do if an accident is unavoidable? Who will it prioritize? How will this be determined [13]?

7.4 AUTONOMOUS TRANSPORT AND SMART CITIES

Autonomous transport stands to revolutionize urban mobility, offering innovative solutions to the challenges posed by growing urbanization. The deployment of autonomous taxis, ride-sharing services, and micro transit options promises to redefine the concept of personal and public transportation [9]. Commuters may experience a shift from traditional car ownership to on-demand autonomous vehicle services possibly resulting in fewer automobiles driving on the road [14].

Consequently, transportation innovations serve as one of the main forces driving the creation of a sustainable transportation network. Coordinated approaches including active transportation, telecommuting, public and shared transportation, and making greater use of the current road capacity are also crucial as they help address important issues like vehicle emissions, transportation costs, travel times, traffic, safety, accessibility, and social equity [3].

The integration of autonomous shuttles and buses into public transit systems holds the potential to enhance accessibility, reduce congestion, and optimize routes. Cities that embrace these technologies may see a transformation in their urban landscape, with implications for parking infrastructure, road design, and public transit planning.

Recent developments are pressuring municipal governors and automakers to adopt a more shrewd and cooperative approach to conducting business. The major automobile companies all see autonomous cars as the way of the future, and they have already made significant progress in their manufacturing processes [15].

The integration of fully autonomous vehicles with smart cities offers the several benefits listed below [9].

1. Improved mobility: integration can result in increased transportation efficiency, decreased traffic congestion, and optimal routing, all of which improve city mobility.
2. Sustainability: the integration attempts to improve cities' sustainability by lowering emissions, energy consumption, and overall environmental effect.
3. Efficient resource consumption: it has the potential to create more synergistic ecosystems that accommodate both city sustainability and efficient mobility for FAGVs, hence maximizing resource utilization.
4. Safety and optimization: real-time data analytics and swarm intelligence may help create safer and more harmonious smart environments, decreasing the strain on cities while making them smarter and more sustainable.
5. Urban dynamics transformation: integration has the potential to revolutionize urban dynamics by allowing for permitting community- and citizen-tailored service planning through participation, bringing citizens and urban dynamics front and center in development.

7.5 CHALLENGES AND FUTURE OUTLOOK OF AUTONOMOUS VEHICLES

While the progress in autonomous land transport is promising, several challenges persist. Technical hurdles, such as the need for improved vehicle sensor reliability in diverse weather conditions and the development of fail-safe mechanisms, remain areas of active research and development [16].

The widespread adoption of smart autonomous transport faces challenges related to the development of comprehensive regulatory frameworks [17]. Governments and international bodies need to establish guidelines to guarantee the responsible and safe integration of self-controlled vehicles on both land and water.

It is no longer necessary for human drivers to display a suitable degree of awareness when should they need to take control of the vehicle due to increased levels of automation. But complicated problems might also occur; one of these is how people interact with and comprehend autonomous systems [18].

Public perception and acceptance represent another challenge, as individuals may harbor reservations about trusting their safety to autonomous systems. Cybersecurity concerns, given the increased connectivity of vehicles, pose a potential threat that necessitates continuous vigilance and advancements in cybersecurity measures. As we rely on digital systems, it is crucial to have strong cybersecurity protections. Preserving the safety and integrity of autonomous transportation networks necessitates safeguarding them against cyber-attacks [19].

7.6 AUTONOMOUS VESSELS ON WATER

Waterway transportation accounts for over 90% of the amount of commercial transit worldwide owing to the benefits of substantial volume and minimal cost. The tremendous growth and increasing ship count in the shipping industry are leading to intelligent and fast ship development. In terms of economies, automation, and efficiency in terms of decision-making and control technology, intelligent ships – also referred to as maritime autonomous surface ships, or MASS – have significant benefits over traditional human ships [20].

Maritime autonomy has led to a significant change in the marine industry, which has a long history founded on naval customs. This paradigm shift is a tidal upheaval that is transforming the dynamics of shipping, logistics, exploration, and environmental monitoring; it is more than just incremental improvement. The following discussion will explore the many facets of maritime autonomy in business, covering its technological foundations, uses, financial implications, and general impact on international trade and sustainability.

The development of autonomous ships ushers a new age in environmentally friendly and efficient shipping. These ships, sometimes referred to as "smart ships," cross the oceans with the least amount of human help thanks to advanced technological devices. Advanced sensors, artificial intelligence (AI)-driven navigation systems, and connectivity technologies are installed on autonomous cargo ships to maximize routes, reduce fuel consumption, and enhance overall operational efficiency.

The effects of autonomous ships on the economy are substantial. Shipping enterprises benefit financially from lower labor costs, increased operational efficiency, and the ability to operate around the clock without crew rest limits. The industry's search for commercially and environmentally feasible solutions is congruent with the promises of reduced carbon emissions, accelerated delivery times, and streamlined worldwide supply networks. Rolls Royce predicts that an unmanned oceangoing vessel will make its appearance in 2035 and that the first remotely operated autonomously oceangoing ship will debut earlier, in 2030 [21].

Technology in the marine sector is advancing quickly, which will eventually result in the commercial use of completely autonomous or remotely operated autonomous ships. The incorporation of self-governing technology into aquatic vehicles signifies a paradigm shift in maritime transportation. Autonomous vessels, from unmanned boats to complex ships, utilize advanced technologies such as AI [22], and navigation systems [4] to navigate the waterways efficiently with precision. Proponents of autonomous vessel technology, like Rolls Royce, Wartsila, and Kongsberg Gruppen, assert that the use of unmanned vessels can save crew operating costs and minimize human error [23].

Autonomous vessels combine sophisticated navigation technologies to travel in water bodies with precision. Global Positioning System (GPS) plays a key role, offering accurate location data for route planning and real-time navigation. Radar and LiDAR systems improve situational awareness by identifying obstacles and other vessels around [24]. Advanced control systems allow these vessels to perform precise maneuvers and adjust to changing environmental conditions. They support navigational planning and collision avoidance, two essential tasks for autonomous vessel navigation [25]. Propulsion solutions for autonomous ships, including electric or hybrid engines, provide effective and environmentally beneficial means of propulsion that lessen pollutants and their negative effects on the environment [26].

Autonomous technology is transforming the maritime industry. Water-based transportation systems are seen as a lucrative endeavor, with autonomy making such systems more commercially feasible [25]. Unmanned cargo ships equipped with state-of-the-art autonomous technology provide several advantages. They have the potential to improve transport routes, minimize fuel consumption, and run independently over long periods, all of which enhance the effectiveness and cost savings of the global shipping network.

Beyond cargo ships, autonomous water vehicles are also utilized in logistics as autonomous surface vessels for research, mapping, and surveying. These boats can navigate challenging terrain and gather vital data for environmental monitoring and oceanographic research. Underwater autonomous vehicles play a crucial role in ocean exploration and research [23]. These remotely operated or autonomous underwater vehicles (ROVs and AUVs), are equipped with advanced sensors and cameras to enable them to explore the depths of the ocean for scientific, commercial, and environmental purposes.

AUVs and ROVs assist in inspecting and maintaining underwater infrastructure in industries like offshore oil and gas, reducing the need for human divers and operational hazards. AUVs equipped with cutting-edge sensors facilitate scientific research, environmental monitoring, and underwater mapping, providing geologists, oceanographers, and marine biologists with invaluable data.

Marine surveying and research are benefiting greatly from the use of unmanned surface vessels (USVs). These ships may collect data for seabed research, bathymetric mapping, and marine environment monitoring as they are equipped with a variety of devices. Because of their autonomy, unmanned surface vehicles can efficiently cover large areas at a lower cost than traditional surveying techniques [27].

USVs are used in environmental studies to investigate marine biodiversity, water quality, and ocean currents. Their capacity to function independently for prolonged durations renders them indispensable instruments for enduring environmental observation and climate study.

As road traffic grows more and more congested, ASVs may offer an enticing substitute for the current transportation system in many coastal and riverbank cities. Relocating some transit from roads to waterways and deploying a self-driving surface fleet could lead to minimal traffic congestion in these water-related communities. [28].

Autonomous water vessels greatly enhance safety and security through applications in search and rescue missions and maritime surveillance. Unmanned devices with sensors and cameras are able to navigate dangerous environments and provide real-time data for monitoring and handling maritime situations [23].

The deployment of autonomous surveillance vessels enhances maritime domain awareness, helping authorities monitor and control maritime activities. These vessels contribute to border security, the protection of critical infrastructure, and the prevention of illegal activities in maritime zones.

7.7 CHALLENGES AND FUTURE PROSPECTS OF AUTONOMOUS VESSELS

Despite the promising developments in autonomous vessels on water, several challenges must be addressed for their widespread adoption. Regulatory frameworks governing autonomous maritime operations need to be established to ensure safety, security, and compliance with international maritime laws.

Technological challenges include enhancing the reliability of communication systems in remote maritime areas and developing robust collision avoidance algorithms [29]. The need for standardized communication protocols and interoperability among different autonomous vessels and systems is crucial for seamless integration into existing maritime ecosystems.

In order to gain regulatory approval, the backing of ship owners, operators, seafarers, and wider public acceptability, remote and autonomous ship operations must be at least as safe as those of already operating boats. While remote and autonomous ships may decrease errors caused by humans, they also can introduce new dangers and change some of the ones that are already present. It will be necessary to investigate these conditions and potential solutions [21].

Public acceptance and awareness, similar to autonomous vehicles on land, are important factors that influence the successful deployment of autonomous vessels. Building trust among stakeholders, including mariners, shipping companies, and the general public, is essential for the acceptance of autonomous technology in maritime operations. Decreased reliance on conventional vehicles, especially autonomous vehicles, may encourage the usage of these vehicles for public transportation [6].

7.8 ENABLING TECHNOLOGIES OF AUTONOMOUS TRANSPORT: PIONEERING INNOVATION ON LAND AND WATER

The evolution of autonomous transport has been fueled by a convergence of cutting-edge technologies that seamlessly integrate to redefine mobility on both land and water. Autonomous transport is a synthesis of advances in artificial intelligence (AI), sensors, communication networks, and more. It includes everything from self-driving automobiles navigating city streets to unmanned boats sailing through waterways [25]. We explore the underlying technologies that support autonomous land and marine transportation, revealing the intricate details of innovation that are transforming the face of transportation in the future.

7.8.1 Artificial intelligence (AI) in autonomous transport

Autonomous land vehicles water vessels rely on artificial intelligence to sense their surroundings, make decisions, and maneuver through intricate situations in real time. Artificial intelligence systems are always learning from their observations and recognizing and comprehending their surroundings through the use of pertinent attributes. Its ability to think like a human and save lives would improve traffic safety and autonomy [30].

AI powered perception systems are essential to giving autonomous cars a thorough awareness of their surroundings. Real-time data is captured by sophisticated sensor arrays, such as LiDAR, radar, cameras, and ultrasonic sensors. After processing this data, AI systems enable cars to identify objects, people, traffic signs, and other important environmental aspects.

Smart autonomous transport systems depend heavily on data collected from various sources, including cameras, sensors, and GPS units, to generate insights and enhance transportation operations. Machine learning (ML) techniques and algorithms are used to process this data [31]. These algorithms enhance the vehicle's capacity to predict potential obstacles, adjust to varying road conditions, and make intelligent decisions by continuously learning from a variety of datasets. Autonomous vehicles are enabled to navigate safely in a variety of dynamic conditions through the iterative learning process [10].

Path planning, action prediction, and obstacle avoidance strategies are employed throughout the decision-making stage to swiftly provide a workable action plan. One of the biggest difficulties drivers encounter when navigating through traffic is reacting to the potential behaviors of other vehicles, which directly affects their own driving approach. This is particularly true at traffic change locations and on roads with several lanes [32].

AI-based decision-making systems play a huge role in autonomous vehicles' capacity to react in intricate situations. These systems analyze data from sensors, interpret traffic conditions, and make decisions on factors like speed, lane changes, and route optimization [30]. Machine learning models enable vehicles to adapt to unforeseen circumstances and learn from real-world experiences.

7.8.2 Sensing technologies in autonomous transport

Autonomous cars and vessels are equipped with sensing technologies that enable them to see and interpret their surroundings accurately. The calculation units receive vital sensing and detection data from sensor systems. Common onboard sensors for autonomous vehicles include cameras, GNSS, LiDAR, radar, ultrasonic, and inertial measurement units (IMUs) [18]. The selection of an appropriate array of sensors and their optimal configurations is a critical component of any autonomous driving system, as it will essentially be used to replicate human perception and judgement in constructing a reliable picture of the environment [16].

Light detection and ranging (LiDAR) are used for localization and mapping applications, essentially giving AVs the ability to avoid obstacles. It determines distance by monitoring how long it takes for a beam to reflect back after reflecting off of objects. Among its many uses are the creation of high-definition maps, obstacle detection ahead of time, and vehicle localization [32].

LiDAR technology maps the environment around a vehicle in three dimensions using laser beams. LiDAR sensors contribute to precise object detection, distance measurement, and mapping, enabling autonomous vehicles to navigate with a high level of accuracy, especially in challenging conditions like low visibility.

Radar sensors identify objects by using radio waves to determine their direction, speed, and distance. Radar is a vital tool for enhancing the perceptive powers of driverless cars, particularly in conditions where visibility is compromised by conditions like dense fog or rain.

The final line of defense for avoiding obstacles is the sonar and radar equipment. The data generated by radar and sonar shows the distance to the nearest item ahead of the vehicle's path. When it detects impediments approaching and that an accident may occur, the autonomous vehicle should engage the brakes or change direction to avoid them. As a result, the data generated by radar and sonar require less processing and are usually sent directly to the control processor, avoiding the main computing pipeline, so that crucial functions like swerving and applying the brakes can be performed [32].

Cameras are fundamental for visual perception in autonomous vehicles. High-resolution cameras capture images and videos of the surroundings, allowing AI algorithms to identify and classify objects, read road signs, and recognize traffic signals. Camera-based vision systems contribute to real-time decision-making and situational awareness.

Current systems often surround the car with eight or more cameras to improve autonomous vehicle safety. Cameras function to identify, track, and detect objects in front of the car, behind it, and to its sides. When combined, these cameras may provide several gigabytes of unprocessed data per second at a frequency of 60 Hz [32].

7.9 CONNECTIVITY AND COMMUNICATION SYSTEMS

The challenge of providing smart network intelligence in self-driving vehicles (AVs) involves creating strong communication networks to handle the intricate data flows necessary for the functioning of AVs. AVs produce and send enormous volumes of data, which need to be processed and evaluated instantly. This data includes sensor, traffic, and mapping data. Complex communication networks, which can manage large data quantities, offer minimal latency and guarantee dependable connectivity even in difficult settings are needed for operations [33].

7.9.1 Vehicle-to-everything communication (V2X)

The interaction of autonomous vehicles with other cars, pedestrians, traffic signals, and other infrastructure elements is facilitated by V2X communication. Cooperative driving is enhanced when automobiles can communicate in both directions and share information about their location, speed, and intentions. Traffic flow is improved, safety is increased, and congestion is reduced with V2X. The ultimate purpose of the upcoming V2X communication systems is to provide cooperative automated driving, reduce collisions, and maximize the amount of space on the road [34].

Typically, autonomous cars are equipped with an array of expensive sensors and processing systems to guarantee the dependability and security of their driving mechanisms. This adds to the overall cost of the vehicle and hinders the widespread use of autonomous vehicles. V2X is a workable substitute for reducing the price of autonomous driving cars since it enables data sharing between automobiles and computation offloading to RSUs [32].

The Internet of Vehicles (IoV) is a network of vehicles and infrastructure that is connected to one another to enable information exchange between different parts of the transportation system. The four main modes of IoV communication are Vehicle-to-Vehicle (V2V), Vehicle-to-Pedestrian (V2P), Vehicle-to-Infrastructure (V2I), and Infrastructure-to-Infrastructure (I2I). V2V communication is the sharing of heading, location, and speed data. Conversely, V2I communication allows cars to communicate with traffic lights, road signs, and other roadside equipment. V2P communication allow vehicles and non-vehicle entities, such scooters and bicycles, to communicate with one another. Finally, data transmission between infrastructure components, such as traffic control systems and weather sensors, is made possible by I2I connectivity. These different communication modalities work together to enable the Internet of Vehicles (IoV) to function as a single, integrated system where infrastructure and vehicles work together to maximize user convenience, efficiency, and safety [31].

7.9.2 5G technology and edge computing

The networking landscape of autonomous transportation has been significantly impacted by the introduction of 5G technology. 5G facilitates real-time data interchange between autonomous vehicles and the cloud by offering incredibly fast and low-latency communication. Thanks to this connectivity, vehicles may communicate with central control systems, view current maps, and get traffic information, all of which improve overall responsiveness and efficiency [35].

Rather than relying solely on centralized cloud computing, edge computing processes data closer to the source. By processing crucial data locally, it enables autonomous vehicles to make decisions more quickly [32]. This reduces latency and enhances the vehicle's ability to respond swiftly to varying road conditions [24].

7.9.3 Blockchain for secure and transparent operations

Blockchain technology is becoming a strong option for guaranteeing safe, open, and effective operations in autonomous land and marine transportation. The technology could have a big impact on how intelligent transportation systems (ITS) are developed and implemented in smart cities. ITS refers to cutting-edge technological applications intended to raise the sustainability, safety, and efficiency of transportation networks [36]. The decentralized and secure platform that blockchain can offer for data storage and exchange can improve the security, privacy, and interoperability of ITS systems. A number of blockchain applications in information and transaction technology (ITS), such as safe data sharing, smart contracts, and decentralized markets, have been proposed recently. Furthermore, the potential of blockchain technology in ITS extends beyond its application in technology. Intelligent transportation systems (ITS) powered by blockchain technology can open possibilities for new businesses to enter the transportation sector and support innovative business models. Blockchain technology can boost trust amongst multiple transportation ecosystem stakeholders in addition to enabling data interchange and administration on an open and secure platform. In conclusion, blockchain technology presents several opportunities for ITS in smart cities [31].

7.9.4 Public health consideration

Autonomous vehicles are a groundbreaking transportation innovation with significant implications for public health. The health impacts, whether positive or negative, depend on factors such as usage patterns, ownership models, propulsion systems, fuel sources, and integration with other transportation modes. Electric AVs, especially when utilized in ridesharing formats and integrated with public and active transport, hold promise for promoting physical activity, improving urban air quality, and creating healthier urban environments. Conversely, AV deployment models focused on individual usage, fossil fuel reliance, increased travel distances, and exacerbated traffic congestion pose risks such as sedentary behavior and worsened urban air quality. Prioritizing research to understand AV market penetration and their effects on travel behavior and urban development is crucial for informing health frameworks and regulations. Recommendations should support policies favoring electric AVs in ridesharing formats, prioritizing public and active transportation, particularly in underserved communities, to foster healthier urban environments [37].

7.9.5 Ethical considerations

Autonomous vehicles and vessels rely on complex decision-making algorithms to navigate and respond to various scenarios. Ethical considerations arise in determining how these algorithms prioritize and make decisions in situations involving potential harm or accidents [38].

First, the data, model, and use of sensitive factors by AV makers and programmers may induce bias into the AVs' algorithms, either purposefully or inadvertently, causing the algorithms to assign higher safety hazards to particular categories of road users. It is difficult to identify algorithmic bias because of the opaque and unpredictable nature of machine learning algorithms, people's overconfidence in algorithms, and the seeming objectivity of algorithmic choices [38].

The collection and utilization significant data volume in autonomous transportation raises security and privacy problems. Safeguarding sensitive information and ensuring responsible data management practices are crucial aspects of addressing these ethical concerns.

7.10 CASE STUDY

Companies today are leading research in the domain of autonomous transportation. Recently, the automotive and maritime industries witnessed remarkable progress in the development of smart autonomous transport systems. Companies like Tesla, Volvo, Rolls Royce, and Roboat are at the forefront of this technological revolution, pushing the boundaries of what is possible on both land and water. Not only are these creative solutions altering the way we travel, but they are also establishing new benchmarks for sustainability, efficiency, and safety.

Elon Musk's Tesla has emerged as a major player in the autonomous driving market thanks to its Autopilot feature. Tesla's Autopilot is an advanced driver-assistance technology that uses a range of sensors, cameras, and artificial intelligence to enable semi-autonomous driving. The system's capabilities include autonomous lane changes, adaptive cruise control, lane-keeping, and parking. Notable among Tesla's Autopilot feature is its over-the-air (OTA) update mechanism, which allows the continual upgrade and expansion of the software.

With its Intellisafe technology, Volvo, a longtime advocate for automotive safety, has welcomed the age of self-driving vehicles. Volvo Intellisafe is a comprehensive suite of safety technologies that include adaptive cruise control, pedestrian identification, automatic emergency braking, and automatic emergency braking with the goal of protecting both pedestrians and vehicle occupants. Intellisafe is designed to seamlessly shift from manual to autonomous driving by integrating with the driver's activities.

While Tesla and Volvo focus on land-based vehicles, a research team based in Amsterdam has developed an autonomous water vessel called Roboat. The boat was designed for a variety of applications, including urban transportation and environmental monitoring.

Roboat can precisely navigate waterways, avoiding obstructions and adjusting to changing conditions because to its sophisticated artificial intelligence-powered navigation system. Roboat demonstrates the potential for autonomous systems to address problems outside conventional road-based solutions with applications ranging from smart city planning to effective transportation on water.

The team considered three tasks in Amsterdam's waterways to showcase the efficacy of the autonomous system: Trajectory Tracking Task, Water Taxi Task and Autonomous Docking. Roboat must complete the Trajectory Tracking Task by following a sinusoidal trajectory at a set speed while carrying different number of passengers. This challenge assesses how well the adaptive NMPC tracking control handles different scenarios for riders. To fulfill the Water Taxi Task, Roboat must operate an autonomous water taxi service, which entails a series of steps including picking up riders at a docking station, maneuvering to a destination, and dropping off the passengers. The autonomy of Roboat is demonstrated in a practical challenge, the Water Taxi challenge. Autonomous Docking demonstrates how Roboat docks into the docking station on its own using its docking algorithm [28].

An innovative ASV, Roboat has been created with the capability to carry up to six people through cities. It can do autonomous docking in urban waterways, obstacle avoidance, course planning, and adaptive receding horizon tracking. When it was tested in busy canals, the goal of Roboat's future development is to incorporate traffic restrictions into the planner. To further enhance tracking effectiveness in noisier environments, incorporate model-based disturbances like waves and currents into its controllers. Robust algorithms will be developed to regulate the formation and self-assembly of several robots in water, facilitating the building of large-scale infrastructure that can be assembled on demand [28].

7.11 CONCLUSION

Looking ahead, the future of autonomous land vehicles and watercraft is dynamic and varied. Auto manufacturers and other technical companies are experimenting with autonomous cars.

Vehicles present greater benefits compared to traditional vehicles, including safety improvement for passengers, pedestrians, and other vehicles. AVs stand to create an arena of new business opportunities, and a considerable shift in transportation convenience for people who cannot or do not want to drive. Several challenges however need to be addressed before commercial autonomous cars are fully deployed on the road. It is expected that continued research and development will solve current problems and provide increasingly sophisticated and reliable autonomous systems. Working together, the auto industry, tech companies, and government organizations will be essential in determining how autonomous land vehicles develop and ultimately transform transportation in the future.

In the same vein, autonomous water transport presents unique opportunities in the maritime industry. Research teams and the industry, are working on increasing the number of unmanned systems, including full-size vessels, in operations at sea. The vessels while revolutionizing mobility at sea by offering benefits such as increased efficiency, reduced crew requirements, and improved safety, also present challenges in regulation and legislation, safety and reliability concerns, protection against cyber-attacks and public perception and acceptance.

REFERNCES

[1] S. S. Canada, "Driving change technology and the future of the automated vehicle. Report of the Standing Senate Committee on Transport and Communications," 2018.

[2] D. Paret, H. Rebaine, and B. A. Engel, "The Buzz about Autonomous and Connected Vehicles," in *Autonomous and Connected Vehicles*, Wiley, pp. 3–22, 2022, doi: 10.1002/9781119816140.ch1.
[3] L. Butler, T. Yigitcanlar, and A. Paz, "Smart Urban Mobility Innovations: A Comprehensive Review and Evaluation," *IEEE Access*, vol. 8., pp. 196034–196049, 2020, doi: 10.1109/ACCESS.2020.3034596.
[4] P. Prasad, "e-Autonomous air, sea, undersea, and land vehicles," *International Journal on Orange Technologies* (IJOT), 2019. [Online]. Available: www.researchparks.org
[5] T. Yigitcanlar, M. Wilson, and M. Kamruzzaman, "Disruptive Impacts of Automated Driving Systems on the Built Environment and Land Use: An Urban Planner's Perspective," *Journal of Open Innovation: Technology, Market, and Complexity*, vol. 5, no. 2, p. 24, Jun. 2019, doi: 10.3390/joitmc5020024
[6] A. Faisal, M. Kamruzzaman, T. Yigitcanlar, and G. Currie, "Understanding Autonomous Vehicles," *Journal of Transport and Land Use*, vol. 12, no. 1, pp. 45–72, 2019, doi: 10.2307/26911258
[7] A. K. Tyagi and S. U. Aswathy, "Autonomous Intelligent Vehicles (AIV): Research Statements, Open Issues, Challenges and Road for Future," *International Journal of Intelligent Networks*, vol. 2, pp. 83–102, Jan. 2021, doi: 10.1016/j.ijin.2021.07.002
[8] S.-C. Chen, "Multimedia for Autonomous Driving," *IEEE MultiMedia*, vol. 26, no. 3, pp. 5–8, Jul. 2019, doi: 10.1109/MMUL.2019.2935397
[9] K. Kuru and W. Khan, "A Framework for the Synergistic Integration of Fully Autonomous Ground Vehicles with Smart City," *IEEE Access*, vol. 9, pp. 923–948, 2021, doi: 10.1109/ACCESS.2020.3046999
[10] R. Hussain and S. Zeadally, "Autonomous Cars: Research Results, Issues, and Future Challenges," *IEEE Communications Surveys and Tutorials*, vol. 21, no. 2, pp. 1275–1313, Apr. 1, 2019, doi: 10.1109/COMST.2018.2869360
[11] "Automated Vehicle Safety | NHTSA." Accessed: Jan. 20, 2024. [Online]. Available: www.nhtsa.gov/vehicle-safety/automated-vehicles-safety
[12] World Health Organization, "Global status report on road safety 2023," 2023. Accessed: Jan. 20, 2024. [Online]. Available: https://iris.who.int/bitstream/handle/10665/375016/9789240086517-eng.pdf?sequence=1
[13] E. Spoehel and S. Banik, "Self Driving Cars: All You Need to Know," in *Proceedings – 2020 International Conference on Computational Science and Computational Intelligence, CSCI 2020*, Institute of Electrical and Electronics Engineers Inc., Dec. 2020, pp. 655–658, doi: 10.1109/CSCI51800.2020.00117
[14] M. Mohammadzadeh, "Sharing or Owning Autonomous Vehicles? Comprehending the Role of Ideology in the Adoption of Autonomous Vehicles in the Society of Automobility," *Transportation Research Interdisciplinary Perspectives*, vol. 9, p. 100294, Mar. 2021, doi: 10.1016/j.trip.2020.100294
[15] B. Padmaja, C. V. K. N. S. N. Moorthy, N. Venkateswarulu, and M. M. Bala, "Exploration of Issues, Challenges and Latest Developments in Autonomous Cars," *Journal of Big Data*, vol. 10, no. 1, Dec. 2023, doi: 10.1186/s40537-023-00701-y

[16] D. J. Yeong, G. Velasco-hernandez, J. Barry, and J. Walsh, "Sensor and Sensor Fusion Technology in Autonomous Vehicles: A Review," *Sensors*, vol. 21, no. 6. MDPI AG, pp. 1–37, Mar. 02, 2021, doi: 10.3390/s21062140

[17] D. Paret, H. Rebaine, and B. A. Engel, "Aspects Relating to Autonomous and Connected Vehicles," in *Autonomous and Connected Vehicles*, Wiley, 2022, pp. 23–80. doi: 10.1002/9781119816140.ch2.

[18] L. Chen *et al.*, "Milestones in Autonomous Driving and Intelligent Vehicles Part I: Control, Computing System Design, Communication, HD Map, Testing, and Human Behaviors," May 2023, doi: 10.1109/TSMC.2023.3276218

[19] D. Tokody, A. Albini, L. Ady, Z. Raynai, and F. Pongrácz, "Safety and Security through the Design of Autonomous Intelligent Vehicle Systems and Intelligent Infrastructure in the Smart City," *Interdisciplinary Description of Complex Systems*, vol. 16, no. 3, pp. 384–396, 2018, doi: 10.7906/indecs.16.3.11

[20] T. Kim and J.-U. Schröder-Hinrichs, "Research Developments and Debates Regarding Maritime Autonomous Surface Ship: Status, Challenges and Perspectives," *New Maritime Business WMU Studies in Maritime Affairs*, Springer, 2021, pp. 175–197, doi: 10.1007/978-3-030-78957-2_10

[21] Rolls Royce, "Autonomous ships The next step." [Online]. Available: www.rolls-royce.com/marine

[22] W. Guan, Z. Cui, and X. Zhang, "Intelligent Smart Marine Autonomous Surface Ship Decision System Based on Improved PPO Algorithm," *Sensors*, vol. 22, no. 15, p. 5732, Jul. 2022, doi: 10.3390/s22155732

[23] Q. Chen, Y. Lau, P. Zhang, M. A. Dulebenets, N. Wang, and T. Wang, "From Concept to Practicality: Unmanned Vessel Research in China," *Heliyon*, vol. 9, no. 4, p. e15182, Apr. 2023, doi: 10.1016/j.heliyon.2023.e15182

[24] S. Baidya, Y.-J. Ku, H. Zhao, J. Zhao, and S. Dey, "Vehicular and Edge Computing for Emerging Connected and Autonomous Vehicle Applications," in *2020 57th ACM/IEEE Design Automation Conference (DAC)*, IEEE, Jul. 2020, pp. 1–6, doi: 10.1109/DAC18072.2020.9218618

[25] Y. Gu and S. W. Wallace, "Operational Benefits of Autonomous Vessels in Logistics—A Case of Autonomous Water-Taxis in Bergen," *Transportation Research Part E: Logistics and Transportation Review*, vol. 154 , Oct. 2021, doi: 10.1016/j.tre.2021.102456

[26] J. L. D. Dantas and G. Theotokatos, "A Framework for the Economic-Environmental Feasibility Assessment of Short-Sea Shipping Autonomous Vessels," *Ocean Engineering*, vol. 279, p. 114420, Jul. 2023, doi: 10.1016/j.oceaneng.2023.114420

[27] F. Sotelo-Torres, L. V. Alvarez, and R. C. Roberts, "An Unmanned Surface Vehicle (USV): Development of an Autonomous Boat with a Sensor Integration System for Bathymetric Surveys," *Sensors*, vol. 23, no. 9, p. 4420, Apr. 2023, doi: 10.3390/s23094420

[28] W. Wang *et al.*, "Roboat III: An Autonomous Surface Vessel for Urban Transportation," *Journal of Field Robotics*, vol. 40, no. 8, pp. 1996–2009, Dec. 2023, doi: 10.1002/rob.22237

[29] H. Lyu *et al.*, "Ship Autonomous Collision-Avoidance Strategies—A Comprehensive Review," *Journal of Marine Science and Engineering*, vol. 11, no. 4. MDPI, Apr. 01, 2023, doi: 10.3390/jmse11040830
[30] N. George and J. Thomas, "Authenticating Communication of Autonomous Vehicles With Artificial Intelligence," *IOP Conference Series: Materials Science and Engineering*, vol. 396, p. 012017, Aug. 2018, doi: 10.1088/1757-899X/396/1/012017
[31] D. Das, S. Banerjee, P. Chatterjee, U. Ghosh, and U. Biswas, "Blockchain for Intelligent Transportation Systems: Applications, Challenges, and Opportunities," *IEEE Internet Things J*, vol. 10, no. 21, pp. 18961–18970, Nov. 2023, doi: 10.1109/JIOT.2023.3277923
[32] S. Liu, L. Liu, J. Tang, B. Yu, Y. Wang, and W. Shi, "Edge Computing for Autonomous Driving: Opportunities and Challenges," *Proceedings of the IEEE*, vol. 107, no. 8, pp. 1697–1716, Aug. 2019, doi: 10.1109/JPROC.2019.2915983
[33] M. Sadaf *et al.*, "Connected and Automated Vehicles: Infrastructure, Applications, Security, Critical Challenges, and Future Aspects," *Technologies (Basel)*, vol. 11, no. 5, p. 117, Sep. 2023, doi: 10.3390/technologies11050117
[34] M. Boban, A. Kousaridas, K. Manolakis, J. Eichinger, and W. Xu, "Connected Roads of the Future: Use Cases, Requirements, and Design Considerations for Vehicle-to-Everything Communications," *IEEE Vehicular Technology Magazine*, vol. 13, no. 3, pp. 110–123, Sep. 2018, doi: 10.1109/MVT.2017.2777259
[35] S. Hakak *et al.*, "Autonomous Vehicles in 5G and Beyond: A Survey," *Vehicular Communications*, vol. 39, p. 100551, Feb. 2023, doi: 10.1016/j.vehcom.2022.100551
[36] X. Wang, S. Garg, H. Lin, G. Kaddoum, J. Hu, and M. M. Hassan, "Heterogeneous Blockchain and AI-Driven Hierarchical Trust Evaluation for 5G-Enabled Intelligent Transportation Systems," *IEEE Transactions on Intelligent Transportation Systems*, pp. 1–10, 2021, doi: 10.1109/TITS.2021.3129417
[37] D. Rojas-Rueda, M. J. Nieuwenhuijsen, H. Khreis, and H. Frumkin, "Autonomous Vehicles and Public Health," *Annual Review of Public Health*, vol. 41, pp. 329–345, 2020, doi: 10.1146/annurev-publhealth
[38] H. S. M. Lim and A. Taeihagh, "Algorithmic Decision-Making in AVs: Understanding Ethical and Technical Concerns for Smart Cities," *Sustainability (Switzerland)*, vol. 11, no. 20, Oct. 2019, doi: 10.3390/su11205791

Chapter 8

Innovation in STEM education

Robotics and AI

Prabhu Jyot Singh and Raj Sandu

8.1 INTRODUCTION: BACKGROUND AND DRIVING FORCES

Artificial intelligence (AI) and robotics have become integral components of science, technology, engineering, and mathematics (STEM) education, revolutionizing traditional learning methods. The current educational approaches often fail to align with industry demands, necessitating a shift towards innovative changes through the incorporation of AI and robotics in education. It not only enhances students' learning capabilities but also transforms their overall educational experience. Smart learning, a cornerstone in the modern curriculum, becomes achievable through these advanced teaching tools, fostering critical thinking, logical reasoning, and real-time practical experiences facilitated by experimental robots.

Robotics and AI technologies play important roles in STEM education. With the help of these technologies, students can acquire the skills required for today's/modern industries, thus improving their chances of securing future jobs. In this chapter, we examine different perspectives on how innovation in STEM education, aided by robotics and AI, completely changes the learning environments for students. To fully comprehend the factors and benefits of students using AI techniques and robotics to enhance their education, it is essential to understand the integration of AI and robotics in STEM education. In the current education system, students cannot be skilled just by reading textbooks, they also require a better understanding of how to use advanced technology in STEM education. Students' skills can be advanced by motivating and guiding them to use AI and robotics' technologies, whether they are studying science, technology, engineering, or mathematics.

Meeting the demands of the fast-paced industry requires students to utilize advanced tools and technologies in STEM education with the help of robotics and AI. These tools empower them to compete effectively in a rapidly changing world. AI and robotic systems create an enriched learning environment by providing educational robots equipped with programmable

 DOI: 10.1201/9781003537083-8

devices and AI-based learning platforms. This transformative approach deviates from traditional learning methods. Educational institutions play a pivotal role in incorporating these techniques and tools into their teaching strategies, offering students foundational platforms to grasp the AI tools essential in today's education systems.

Ongoing research actively explores effective STEM education methods to align with industry demands. Xu and Ouyang [1] delve into the application of AI technology in STEM education, highlighting key AI applications such as automation, educational robots, intelligent tutoring systems, learning prediction, and discussions on student behavior. Kong et al. [2] contribute insights into the role of AI in STEM education, emphasizing three distinct aspects: the effective use of AI for STEM literacy, the contribution of AI activities to the enjoyment of STEM knowledge, and the interconnectedness between AI elements and STEM activities in a language accessible to human readers.

The remainder of this chapter unfolds as follows: Section 8.2 explores the foundation of STEM education, while Sections 8.3 and 8.4 discuss robotics and AI applications in STEM education, respectively. Pedagogical approaches in STEM education are covered in Section 8.5, with Section 8.6 focusing on curriculum integration. Section 8.7 addresses technology and resource requirements, and real-world applications and benefits and challenges of robotics and AI in STEM education are detailed in Sections 8.8 and 8.9, respectively. The chapter concludes in Section 8.10, summarizing the key findings and insights.

8.2 THE FOUNDATIONS OF STEM EDUCATION

In today's world, education plays a crucial role in the success of every individual. The education domain offers various fields that enable students to choose their path and specialize in a particular area, with STEM education being one of them. Science, technology, engineering, and mathematics are four distinct disciplines from which students can choose their future careers. A pertinent question arises regarding whether studying these disciplines alone is sufficient for a student to thrive in the rapidly evolving global market.

These four STEM disciplines are interconnected, and delving into the study of these interrelated fields equip students with greater knowledge and practical skills in addressing complex real-world problems. The pivotal question that requires addressing pertains to the foundation of STEM education. Several key factors contribute to establishing the foundation of STEM education, as depicted in Figure 8.1, providing an overview of these foundational elements. This section will systematically explore each of these factors.

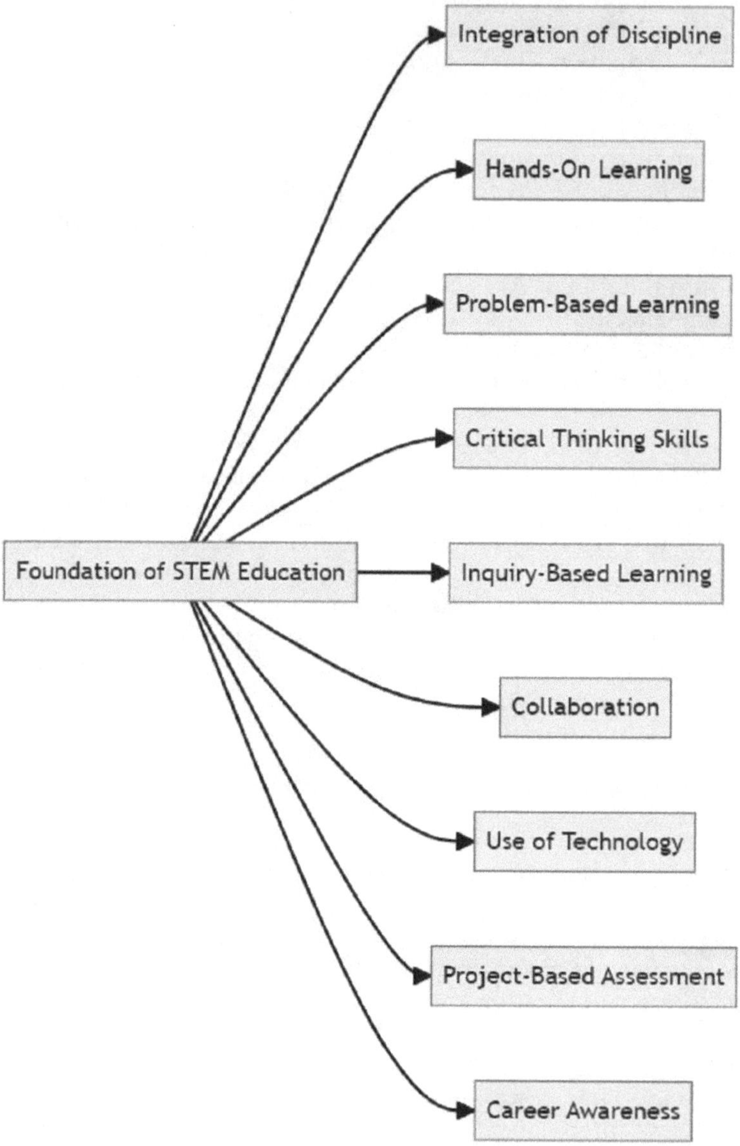

Figure 8.1 Foundation of STEM education.

8.2.1 Integration of discipline

The primary element that establishes a foundation in education within STEM education is the integration of disciplines. Rather than solely studying each independent discipline, students should engage in a specially crafted education plan that integrates science, technology, engineering, and mathematics. The interconnected education of STEM disciplines presents students with a more promising opportunity to become successful individuals. In [3], English explores STEM education in K-12 and emphasizes the significance of integration. The discussion revolves around how integration serves as a crucial factor in STEM education, emphasizing that it should align with the learning outcomes of the students.

8.2.2 Hands-on learning

In contemporary education, students should not solely depend on theoretical concepts; practical, hands-on learning is also essential in STEM education to effectively address real-world complex problems. Yannier et al. [4], introduced an AI-based system that demonstrates how hands-on learning can be utilized to teach STEM education.

8.2.3 Problem-based learning

Problem-based learning forms the cornerstone of STEM education, where students engage with real-world problems and challenges, applying their STEM knowledge to find solutions. The set of principles governing the problem-based learning model in STEM education is elucidated in [5] that investigated crucial principles for STEM education such as flexible knowledge, skills, and capabilities; active and strategic metacognitive reasoning; collaboration driven by intrinsic motivation; and problems embedded in real and rich contexts.

8.2.4 Critical thinking skills

Through STEM education concepts, students analyze problems and apply their critical thinking skills, ultimately concluding to address real-world complex problems. Mutakinati et al. [6] conducted a study with middle school students to assess their critical thinking skills.

8.2.5 Inquiry-based learning

STEM education additionally promotes inquiry-based learning, aligning with student's inherent curiosity to solve critical problems. This involves asking questions, conducting research on relevant topics, and performing investigations to find answers to those questions. Sutaphan and Yuenyong [7] presented a theoretical framework for the STEM education teaching approach, emphasizing inquiry within the context of context-based learning.

8.2.6 Collaboration

STEM education also promotes the collaboration of interdisciplinary students to work on a project where a combination of all four fields of knowledge is required. This collaborative experience allows the student to share their ideas and work as a team for real-world tasks. The advantage of collaboration in STEM education is discussed in [8] where Pasani and Amelia examined developing technology-based collaborative skills through the STEM approach to reflect the trend of integrating technology into education.

8.2.7 Use of technology

We cannot think about education nowadays without involving computer systems such as hardware, software, and applications. Li et al. [9] investigated the research and trends in STEM education and they studied how the use of technology becomes a main foundation in STEM education. Students leverage various technologies and tools in their STEM education study, contributing significantly to time efficiency in grasping key concepts.

8.2.8 Project-based assessment

Project-based assessment learning stands out as another fundamental component in STEM education. The theoretical knowledge acquired by students during their STEM education studies can be actively applied to address complex problems within project-based assessments. In [10], Keleman examined STEM integration learning methods through project-based learning, emphasizing how this approach can enhance student's higher-order thinking skills.

8.2.9 Career awareness

The current education system empowers students to align their career choices with the demands of industries. Within STEM education, students have the opportunity to select from promising careers that meet the current and future needs of various industries.

8.3 ROBOTICS APPLICATIONS IN STEM EDUCATION

Recent technological advancements and tools have significantly enhanced student's comprehension of STEM education, with robotics being a

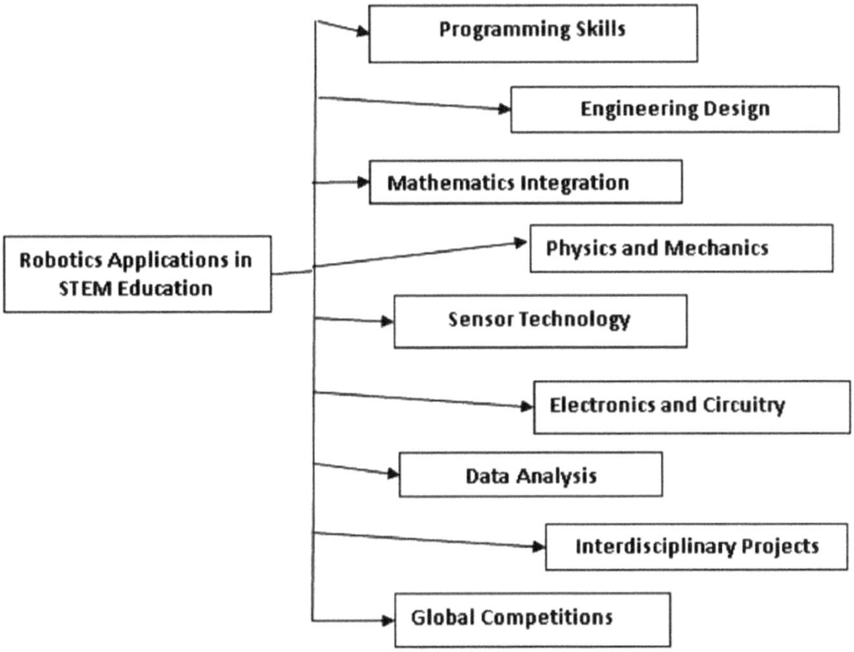

Figure 8.2 Robotics' applications in STEM education.

noteworthy example. Numerous research studies have been conducted in the field of educational robots, shedding light on their applicability in STEM education. In [11] and [12], authors delve into discussions on educational robots. Furthermore, [13] focuses on educational robotics in the secondary education stage, while [14] explores the features of educational robotics and STEM research in primary education.

Robotics' applications play a crucial role in enhancing students' learning experience in STEM education. Figure 8.2 illustrates robotics' applications in STEM education. In this section, we will delve into a detailed discussion of the most widely utilized robotics' applications in STEM.

8.3.1 Programming skills

Robots are used in different industry application domains, and they are required to train through a programming language. Students can learn various programming languages in STEM education and that language can be used to provide instruction to robots to perform a specific task.

8.3.2 Engineering design

Another application of robotics is engineering design. STEM education allows students to learn about engineering design. This knowledge of engineering design can be used by students to design and build robots for various industries.

8.3.3 Mathematics integration

Knowledge of mathematics is also required for building robots for future use. Students learn from mathematics the knowledge of angle, distance, 2D and 3D problems. These concepts help in robot-building projects such as pathfinding and distance-finding applications.

8.3.4 Physics and mechanics

Robots can perform many tasks that are directly related to the movement of robots. These robotic movements can be properly managed and controlled in robots with the knowledge of physics and mechanics concepts such as knowledge about forces and motions.

8.3.5 Sensor technology

The robots that are currently serving in many industries are capable of sensor technology. Students having advanced knowledge about sensor technology such as touch sensors, infrared, and ultrasound will be very helpful for them in building the world's promising robots.

8.3.6 Electronics and circuitry

The electrical engineering concepts such as the interconnection of various components in microcontrollers used in the designing of robots are very helpful. The students who are getting this knowledge through STEM education can solve electronics and circuitry problems in robots.

8.3.7 Data analysis

As we discussed sensor technology earlier, sensors are key components in robots for data collection. The data collected by the robot's sensor needs to be analyzed through which some decision-making conclusions can be determined. STEM education provides the opportunity for students to learn these concepts.

8.3.8 Interdisciplinary projects

The design and implementation of robots involve multi-discipline projects such as engineering knowledge for the design purpose, technology knowledge for the sensor technology, and physics knowledge for the movement. STEM education guides the student to get this overall knowledge of robotics and AI technology.

8.3.9 Global competitions

There are lots of robotics' competitions conducted worldwide every year by big organizations. The knowledge of STEM education helps students to show their capabilities and skills about robotics and their application.

The field of robotics in STEM education is dynamic and continuously evolving. The applications discussed so far are significant, but ongoing research continues to explore and unveil more advanced robotics applications. As technology advances, new possibilities, and innovations in robotics within STEM education are likely to emerge, offering students even more opportunities to engage with cutting-edge concepts and contribute to the ever-growing field of robotics. STEM education remains at the forefront of preparing students for the challenges and advancements in this dynamic and interdisciplinary field.

8.4 AI APPLICATIONS IN STEM EDUCATION

AI stands out as one of the rapidly advancing trends in technology, witnessing extensive research and development in recent years. Its integration with STEM education has resulted in numerous applications. In recent research, several authors have explored the utilization of AI in STEM education. Yannier et al. [4] delve into the discussion of a mixed-reality AI system designed to support STEM education. The incorporation of AI-thinking in science, technology, engineering, arts, and mathematics (STEAM) education is investigated in [15]. A study conducted in [16] offers insights through a systematic mapping review of augmented reality applications supporting STEM learning in higher education. The exploration of generative AI in education and research, along with associated opportunities, concerns, and solutions, is presented in [17]. Additionally, [18] examined the educational applications of the ChatGPT AI system. Figure 8.3 visually represents all these AI applications in STEM education, and we will comprehensively discuss each of these applications in this section.

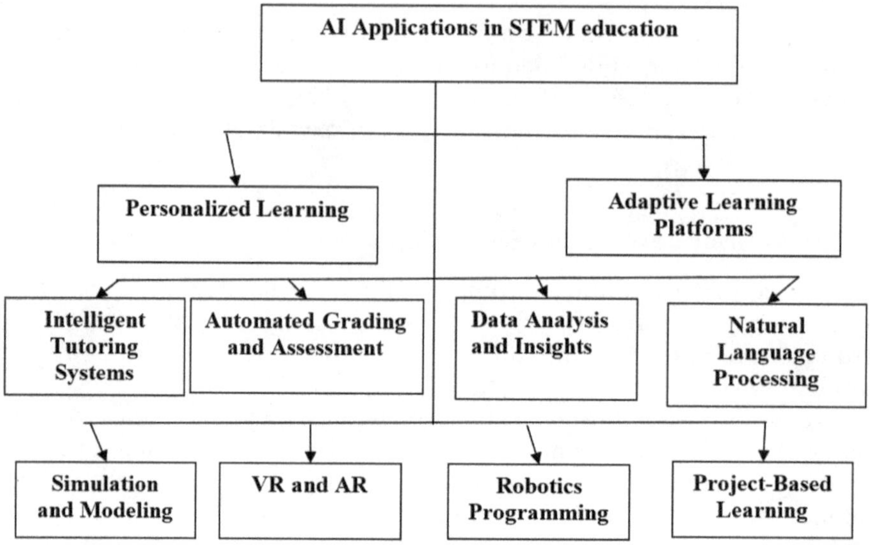

Figure 8.3 AI applications in STEM education.

8.4.1 Personalized learning

With the help of the AI algorithm, a personalized study plan can be prepared for the student who is studying STEM education. This personalized study plan will be based on the study pattern of individual students and will focus on the area where that student is currently studying. This AI algorithm can create personalized learning through the learning model which accepts the student learning pattern as an input and gives the required personalized learning plan.

8.4.2 Adaptive learning platforms

AI platforms can be used for adaptive learning. A student can learn a specific lesson based on their intelligence and knowledge. The student performance can be better evaluated based on the individual student intelligence and knowledge with the help of an AI platform.

8.4.3 Intelligent tutoring systems

Nowadays, students prefer personal tutors to study and discuss various key concepts that are being taught in STEM education. It is a very hard task for each student to get a personal tutor to solve their problems. The AI virtual tutors are the most promising solution to this problem. The virtual tutor can interact with students by answering their questions, explaining the concept, and also making a personalized learning plan.

8.4.4 Automated grading and assessment

AI is also used in grading the student for their assessment. With the help of an AI tool automated grading and feedback can be provided to each student. It is a very time-saving approach for teachers while dealing with the assessment.

8.4.5 Data analysis and insights

STEM education is not only about teaching the STEM concept to students but also the proper designing of curriculum and teaching strategy. AI algorithms can analyze a large amount of data to determine the pattern of study for students who are studying STEM education. This AI data analysis tool is very helpful for teachers to design the curriculum and strategy as per the needs of students.

8.4.6 Natural language processing (NLP)

The AI tool for NLP in STEM education helps students to interact with human-based language. With the help of chatbots and another interactive platform of learning, a student can get the answers to all questions. Students can be more engaged in their studies if the education can be delivered in their natural language.

8.4.7 Simulation and modeling

STEM education is not a theoretical education. It is a combination of theory and practical education. Sometimes students need real-world scenarios and experimental equipment to do some practical work and that is not always possible. The AI simulation tool allows students to practice all the experiments in simulation environments, which helps them to better understand the key concepts of STEM education.

8.4.8 Virtual reality (VR) and augmented reality (AR)

The AI tool for VR and AR provides a virtual laboratory experience to STEM students to get a better understanding of complex concepts. Students can visualize complex concepts with the help of VR and AR technology, which helps them to learn those concepts easily.

8.4.9 Robotics programming and AI integration

Another important AI application is the learning of programming languages and the use of those programming languages to instruct robots. Various AI tools can help STEM students learn different programming languages as per the industry demand. The knowledge of programming language is also useful in robotics where the programming languages are used to provide specific instruction to robots.

8.4.10 Project-based learning support

The AI tool also helps STEM students get project-based learning support by providing the resources and feedback that are required to complete those projects.

While we have covered various AI applications in STEM education in this section, it is important to note that the applications discussed are not exhaustive. The dynamic nature of technology and education means that new AI applications and innovations will likely continue to emerge, further enriching the learning experience for STEM students. The landscape of STEM education is ever-evolving, and AI serves as a catalyst in shaping the future of how students engage with and learn from these disciplines.

8.5 PEDAGOGICAL APPROACHES IN STEM EDUCATION

The question of how to effectively teach students in STEM education is pivotal, especially considering that the traditional memorization-based learning approach falls short in the context of advanced technology-based subjects. The innovation brought by AI and robotics opens up new possibilities for a pedagogical approach to STEM education. These technologies offer students an enriched learning experience, enabling them to grasp core concepts that can be applied to solve future problems.

Recent research in STEM education explores various pedagogical approaches. In [19], Wells introduced the PIRPOSAL model, a conceptual and pedagogical framework for integrative STEM education. This model consists of phases such as problem identification, ideation, research, potential solution, optimization, solution evaluation, alterations, and learned outcome. In [20], the researcher discusses a transformation model of education for the integration of the STEM approach in a pedagogical university, preparing educators for a new formation with indicators of effectiveness. Another study completed in [21] explores an Internet-of-Things curriculum, pedagogy, and assessment for STEM education.

Many industry projects require teamwork, and STEM students are gaining valuable experience in group collaboration to solve specific problems using AI and robotics tools. For instance, simulation tools are employed not just to address theoretical challenges, but also to collaboratively find real-world solutions. This hands-on, real-world problem-solving enhances students' overall learning experience. Internships in industries, often included in STEM curriculum courses, provide students with the opportunity to work on real-world problems using AI tools and collaborate with others.

Ethical considerations and diverse perspective learning are also important approaches for STEM students to enhance their learning experience by understanding the potential bias. The advanced technology of AI and robotics allows STEM students to obtain hands-on experience to get familiar with these tools and also enhance their knowledge for virtual collaboration that are the most promising technology for future generations. The teaching strategy for STEM students should always guide them in their lifelong learning skills with the emerging technology of robotics and AI. Education of these new technologies also allows students to easily move on to any new technology that will come in the future and will be more beneficial for the STEM education students.

8.6 INTEGRATING ROBOTICS AND AI INTO THE CURRICULUM

The integration of robotics and AI in the current STEM education curriculum will prepare students to learn the advanced level concepts that can be helpful for them to complete in future challenges. This new technology's integration in the current STEM education curriculum not only enhances the learning skills and knowledge of students but also prepares them to solve real-world problems. There are lots of factors that should be considered while integrating robotics and AI into the STEM curriculum. In this section, we will analyze those factors in detail.

The first and most important factor for this integration is related to the foundation of STEM education study. If a student does not have a strong foundation in STEM education then it will be hard to explain the new curriculum to them, which includes the integration of robotics and AI technology. For example, if students do not have knowledge of basic programming skills, how we can expect them to perform well with a new curriculum with an advanced level of programming for AI and robotics? The new STEM education curriculum with the integration of AI and robotics should be designed keeping in mind the age of students as well. There should be more practical hands-on activities for younger students and more in-depth knowledge of complex activities for students as they progress in their STEM education journey. The exercise to build a robot and program robots should also be included in the new curriculum, which will help students to understand how this new technology works for real-world problems.

STEM education is an interdisciplinary education and its new curriculum with the integration of AI and robotics should also follow the interdisciplinary education. The various ways of teaching should be used for interdisciplinary subjects with the help of robots and AI. For example, robots can be used to teach different discipline concepts and AI can be used with programming to discuss overall knowledge. The AI and robotics-based STEM education curriculum should be incorporated by coding and programming knowledge. With the help of specialized AI coding tools learning about the programming concept can be a very easy task for students. This new AI and robotics-integrated STEM education curriculum should also include project-based learning, collaborative learning, and the ethical and social implications that will enhance the overall skills of the students. Another issue in this new curriculum is access to resources. For example, the schools that will adopt this new curriculum should have access to robotics' kits, computers, and software as well as support and mentorship from industries and universities. The university expert educator should also guide and motivate all entry-level STEM education students from time to time who are studying this new curriculum. Educators should also enroll in advanced training for AI and robotics to update their knowledge and help them teach students the new STEM education curriculum with the integration of AI and robotics effectively.

Some research has already been conducted to study the effect of AI and robotics' integration in the STEM education curriculum. In [22], Leoste et al. analyzed the strengths and weaknesses of the teaching staff of higher education who are adopting AI and robotics-integrated curricula. The impact of this integration is also studied at school-level education. Lee and Perret [23] conducted a study about preparing high school teachers to adapt the AI method to STEM classrooms.

8.7 TECHNOLOGY AND RESOURCE REQUIREMENTS FOR AI AND ROBOTICS' INTEGRATION

The integration of AI and robotics in STEM education requires technology and resources and that should be carefully considered while using this technology in the new STEM education curriculum. A study conducted in [24, 25], investigated the resources and technology requirements to use AI and robotics in the STEM education classroom. Pedro et al. mentioned in [26] the challenges and opportunities of AI in education and one of the challenges about the requirement for resources. In this section, we analyze the most important technologies and resources that are required for AI and robotics in STEM education.

The first component of the resource requirement for AI and robotics is the hardware, which can be robots or computer devices. Students can start their basic learning with simple programming robots and advance their skills with advanced robotic kits. Computer devices such as desktop computers or tablets are also required for controlling and programming robots. The next resource component that is required for STEM education with the integration of AI and robotics is the software. Students are required to access the programming environment where they can write the code to instruct the robots. The programming environment requirement will be different for young age entry-level students as compared to the older students who can use advanced programming environments such as Python and Java. As the student will be getting access to the hardware robots for practical learning they also require simulation software, which can be used to enhance their practical skills for the learning and development in STEM education. There is also a requirement for software that can be used to teach AI such as Google's AIY Projects or AI education platforms like TensorFlow for Education. Apart from that students also require software and application programs to access the online learning platform and curriculum resource access.

Another important resource for AI and robotics' integration in STEM education is the use of internet technology. Nowadays we cannot think about any education without the use of the internet, the same applies to STEM education. Students who are learning AI and robotics' integrated STEM education need reliable internet connections to access the resources. The advanced level of programming is now based on the cloud platform and cannot be run without the internet as such it is a really important resource for this new technology.

The university students studying the new STEM education curriculum integrated with AI and robotics are also required to access resources such as sensors and actuators for robotics' projects as well as some collaborative platforms where they can share their work with their peers. While students use this new technology and tools with the knowledge of hardware, software, the internet, and another teaching platform they are also required to get access to the documentation about how to use these resources.

8.8 REAL-WORLD APPLICATIONS STEM EDUCATION – ROBOTICS AND AI

Learning about robotics and AI in STEM education opens up numerous practical possibilities. It equips students to pursue careers in fields with tangible real-world impacts and hone their critical thinking and problem-solving abilities. In the following part, we will explore some actual uses of STEM education in the realm of robotics and AI.

8.8.1 Manufacturing and automation

In the realm of STEM education, students delve into manufacturing and automation applications. This involves understanding the intricacies of the manufacturing automation process, encompassing operations like assembly line functioning, design processes, welding procedures, and the execution of picking and packaging tasks aided by industrial robots [27].

8.8.2 Healthcare

Medical robots play a vital role in instructing students on surgery, rehabilitation, and patient care. They offer an effective means of educating medical students, providing hands-on experience in a controlled environment [28].

8.8.3 Agriculture

Students get knowledge about precision farming with the help of autonomous vehicles, drones, and robots, which can be used for monitoring crops, applying fertilizer to crops as per requirement, and also helping in irrigation [29, 30].

8.8.4 Space exploration

Space robots such as rovers and robotics' arms are useful tools for space exploration. Students get a better understanding of space exploration with the help of these space robots [31].

8.8.5 Education and research

Student programming skills can be better with the help of robotics' kits and AI tools that can be used for teaching the program skills to students [32, 33].

8.8.6 Transportation

Self-driving cars and autonomous vehicle robotics can be very helpful in transportation. Learning about real-world transportation robots will help students to better understand robotics' applications in transportation [34, 35].

8.8.7 Environmental monitoring

Underwater drones and aerial drones are very promising to monitor the environment. Students who are studying environmental monitoring can enhance their knowledge by exploring the working functionality of underwater and aerial drones [36, 37].

8.8.9 Finance and business

Students get a better understanding of finance and business with the help of AI-based trading algorithms. Robotics' applications such as Robo Advisor also enhance their knowledge about finance and business [38].

8.8.10 Entertainment and arts

Real-world robotics' applications such as animatronics and the use of robotics cameras change the direction of entertainment and arts. A student who is interested in this area should get knowledge about this application that will add more value to their knowledge [39].

8.8.11 Energy

Robots are being used in the inspection of energy plants to identify faults in the pipeline and powerplant. This real-world application of robotics will help STEM students understand how these robots perfectly fit in the energy section application [40].

8.8.12 Defense and security

The autonomous unmanned aerial vehicle that is being used for rocket launching and the drones that are used for surveillance purposes are the real-world applications of robotics and AI in the defense and security sector. STEM students should gain knowledge of robotics in this area to find out the real-world challenges in this domain [41, 42].

8.8.13 Retail and logistics

Warehouse automation with the help of robotics is the best example of the use of robotics in retail and logistics. The robotics will help in the automation process for packing, packaging ordering, dispatching, and transport operations in a warehouse. STEM education students who are studying retail and logistics can enhance their knowledge by understanding retail and logistic robotics' applications [43, 44].

8.8.14 Cybersecurity

With the help of AI algorithms and tools, a large amount of data can be analyzed to find any data breaches. AI tools are also used to prevent and detect any security threats. The understanding of cyber security issues with the help of AI gives a deep understanding to the STEM student about the security threats, risks, and how countermeasures can be applied to those issues [45, 46].

8.9 BENEFITS AND CHALLENGES OF ROBOTICS AND AI IN STEM EDUCATION

As we discussed in the previous sections, there are lots of benefits of robotics and AI in STEM education such as students will get a better understanding of theoretical as well as practical knowledge in STEM education with the help of advanced technology that is now being used in the industries. At the same time, these technologies also pose some challenges. In this section, we summarize the key benefits and challenges of this technology in STEM education and it is presented in Table 8.1.

Table 8.1 Benefits and challenges of robotics and AI in STEM education

Benefits	*Challenges*
Hands-On Learning: Students can apply the theoretical STEM education knowledge to enhance their understanding with the practical hands-on learning experiment.	**Access to Resources:** Problem with access to resources to institutions such as advanced AI tools and robotic kits.
Problem-Solving Skills: Student working with robotics and AI improves their critical thinking and problem-solving skills.	**Teacher Training:** There are challenges for proper training for educators to use these advanced technologies in STEM education.
Interdisciplinary Learning: In STEM education, students delve into interdisciplinary subjects, acquiring knowledge across various scientific and technical disciplines. This approach proves invaluable as it equips students with a broad skill set, enabling them to tackle complex problems by drawing upon insights from multiple fields.	**Cost Implications:** Cost stands out as a significant challenge in the implementation of AI and robotics in STEM education. The expense associated with acquiring equipment, software, and resources for these technologies can be prohibitive, limiting the accessibility of advanced tools and hindering the broader integration of AI and robotics in educational settings.
Real-World Applications: Robotics and AI knowledge help students choose a future promising career to solve real-world problems using STEM education.	**Curricular Integration:** It is not an easy task to integrate AI and robotics into the STEM education curriculum.
Programming and Coding Proficiency: In this digital world, programming is a very important skill. Students can learn and apply their programming skills very effectively with the help of these technologies.	**Societal Perceptions:** Some of the public perceptions about these technologies are that there will be no jobs in these areas in the future, making it hard for students to choose these courses.
Promotion of Creativity: Student creativity and application of their innovative ideas to build and design robots will help them to deliver their thoughts and improve their creativity.	**Diversity and Inclusion:** Access to equipment and resources for AI and robotic STEM education for students with diverse backgrounds is another big challenge.

Table 8.1 (Cont.)

Benefits	*Challenges*
Preparation for Future Careers: Robotics and AI with STEM education prepare students for future careers that are the most demanding areas in the industries.	**Rapid Technological Advances:** The fast changes in AI and robotics' technologies can make it a problem for educational institutes to keep their curriculum up to date at all times.
Increased Engagement: This new technology, such as working with robots always creates interest among students and increases their engagement to learn about these technologies.	**Assessment Methods:** The assessment of student skills and knowledge for robotics and AI STEM education students is another challenge as the traditional assessment method cannot be used to evaluate the skills of AI and robotics' students.
Global Collaboration: AI and robotics' projects are the most demandable work across the globe. Students studying these subjects will get a chance to collaborate with their peers all around the world with diverse backgrounds, which will help them to build their teamwork skills.	**Limited Interdisciplinary Integration:** It is not an easy task for every educational institute to integrate robotics and AI into their interdisciplinary STEM education.
Ethical Considerations: Robotics and AI STEM education not only equip students with technical skills but also foster awareness of the ethical use of technology and its social impact. Students learn to navigate the ethical considerations and social implications of deploying these advanced technologies, promoting responsible and mindful use. This broader perspective helps develop a generation of professionals who are not only technically proficient but also ethically conscious and socially responsible in their application of robotics and AI.	**Ethical and Privacy Concerns:** Students studying AI and robotics in STEM education often express concerns about the ethical use of data and resources. This reflects a growing awareness of the importance of ethical considerations in the development and deployment of technologies. Addressing these concerns involves instilling a strong ethical framework in the educational curriculum, promoting responsible practices in handling data, and encouraging students to consider the broader ethical implications of their work in AI and robotics.

8.10 CONCLUSION

As technologies are fast-moving and changing year by year it is also required to adapt to those changes in STEM education as well. The use of the innovative development of the emerging technologies, AI and robotics in STEM education opened up a better career path for students. They can get a better understanding of STEM education with the help of these technologies. Students can be able to solve complex problems with the help of advanced learning skills through this new technology and also apply their learning skills to real-world industrial problems.

In this chapter, we covered various aspects of robotics and AI in STEM education. We discussed the foundation of STEM education followed by the application of robotics and AI in STEM education. The pedagogical approach in STEM education and the integration of AI and robotics in the STEM education curriculum is also addressed in this chapter. The resource requirement for AI and robotics in STEM education is also mentioned in this chapter. We also explained the real-world various domain AI and robotics' applications in STEM education in this chapter followed by the benefits and challenges. In this book chapter, we tried to cover all key concepts to understand the innovation of robotics and AI in STEM education.

Many education institutes are adopting this integration of AI and robotics in STEM curriculum. More research needs to be done to show the benefits of using these technologies in STEM education. Many researchers and industries are conducting research in this area to find out how these technologies can better serve STEM education. In the future, we can imagine most of the educational institutes will be providing AI and robotics' study to the students with the integration of STEM education.

REFERENCES

[1] W. Xu and F. Ouyang, "The Application of AI Technologies in STEM Education: A Systematic Review from 2011 to 2021," *International Journal of STEM Education*, vol. 9, no. 1, Sep. 2022, doi: https://doi.org/10.1186/s40594-022-00377-5

[2] S. Kong, H. Ogata, J.-L. Shih, and G. Biswas, "The Role of Artificial Intelligence in STEM Education." Available: https://icce2021.apsce.net/wp-content/uploads/2021/12/ICCE2021-Vol.II-PP.-774-776.pdf

[3] L. D. English, "STEM Education K-12: Perspectives on Integration," *International Journal of STEM Education*, vol. 3, no. 1, Mar. 2016, doi: https://doi.org/10.1186/s40594-016-0036-1

[4] N. Yannier, S. E. Hudson, and K. R. Koedinger, "Active Learning Is About More Than Hands-On: A Mixed-Reality AI System to Support STEM Education," *International Journal of Artificial Intelligence in Education*, vol. 30, no. 1, pp. 74–96, Feb. 2020, doi: https://doi.org/10.1007/s40593-020-00194-3

[5] K. Smith et al., "Principles of Problem-Based Learning (PBL) in STEM Education: Using Expert Wisdom and Research to Frame Educational Practice," *Education Sciences*, vol. 12, no. 10, p. 728, Oct. 2022, doi: https://doi.org/10.3390/educsci12100728

[6] L. Mutakinati, I. Anwari, and Y. Kumano, "Analysis of Students' Critical Thinking Skill of Middle School through STEM Education Project-Based Learning," *Jurnal Pendidikan IPA Indonesia*, vol. 7, no. 1, pp. 54–65, Apr. 2018, doi: https://doi.org/10.15294/jpii.v7i1.10495

[7] S. Sutaphan, and C. Yuenyong, "STEM Education Teaching Approach: Inquiry from the Context Based" *Journal of Physics, Conference Series IOP Publishing*, vol. 1340, no. 1, p. 012003, Oct. 2019.

[8] C. F. Pasani and R. Amelia, Developing Collaborative Skills through STEM Approach. *IntechOpen*, 2023. Accessed: Jan. 05, 2024. [Online]. Available: www.intechopen.com/online-first/88696

[9] Y. Li, K. Wang, Y. Xiao, and J. E. Froyd, "Research and Trends in STEM Education: A Systematic Review of Journal Publications," *International Journal of STEM Education*, vol. 7, no. 1, Mar. 2020, doi: https://doi.org/10.1186/s40594-020-00207-6

[10] M. Keleman, "Assessment of Higher Order Thinking Skills Through Stem Integration Project-Based Learning for Elementary Level," *International Journal of Social Science and Human Research*, vol. 04, no. 04, Apr. 2021, doi: https://doi.org/10.47191/ijsshr/v4-i4-40

[11] T. Sapounidis, and D. Alimisis, "Educational Robotics for STEM: A Review of Technologies and Some Educational Considerations," In *Science and Mathematics Education for 21st Century Citizens: Challenges and Ways Forward*, Hauppauge, NY, USA: Nova Science Publishers, pp. 167–190, Sept. 2020.

[12] A. Chatzopoulos, M. Papoutsidakis, M. Kalogiannakis, and S. Psycharis, "Innovative Robot for Educational Robotics and STEM," *Lecture Notes in Computer Science*, pp. 95–104, Jan. 2020, doi: https://doi.org/10.1007/978-3-030-49663-0_13

[13] N. Arís and L. Orcos, "Educational Robotics in the Stage of Secondary Education: Empirical Study on Motivation and STEM Skills," *Education Sciences*, vol. 9, no. 2, p. 73, Apr. 2019, doi: https://doi.org/10.3390/educsci9020073

[14] S. Tselegkaridis and T. Sapounidis, "Exploring the Features of Educational Robotics and STEM Research in Primary Education: A Systematic Literature Review," *Education Sciences*, vol. 12, no. 5, p. 305, Apr. 2022, doi: https://doi.org/10.3390/educsci12050305

[15] M.-L. How and W. L. D. Hung, "Educing AI-Thinking in Science, Technology, Engineering, Arts, and Mathematics (STEAM) Education," *Education Sciences*, vol. 9, no. 3, p. 184, Jul. 2019, doi: https://doi.org/10.3390/educsci9030184

[16] S. Mystakidis, A. Christopoulos, and N. Pellas, "A Systematic Mapping Review of Augmented Reality Applications to Support STEM Learning in Higher Education," *Education and Information Technologies*, Aug. 2021, doi: https://doi.org/10.1007/s10639-021-10682-1

[17] E. Alasadi and C. R. Baiz, "Generative AI in Education and Research: Opportunities, Concerns, and Solutions," *Journal of Chemical Education*, vol. 100, no. 8, pp. 2965–2971, Jul. 2023, doi: https://doi.org/10.1021/acs.jchemed.3c00323

[18] M. Kallogiannakis, S. Papadakis, A. I. C. Gözüm, and Z. H. İpek, "Educational Applications of the ChatGPT AI System: A Systematic Review Research," *Educational Process: International Journal (EDUPIJ)*, vol. 12, no. 3, pp. 26–55, 2023.

[19] J. Wells, "PIRPOSAL model of integrative STEM education: The premise of I-STEM ED requires a blended approach intent on teaching other expressly targeted STEM content and practices." 2016.
[20] Balyk, G. Shmyger, Y. Vasylenko, V. Oleksiuk, and A. Skaskiv, "STEM-Approach to the Transformation of Pedagogical Education," *dspace.tnpu.edu.ua,* 2019.
[21] P. Abichandani, V. Sivakumar, D. Lobo, C. Iaboni, and P. Shekhar, *"Internet-of-Things Curriculum, Pedagogy, and Assessment for STEM Education: A Review of Literature | IEEE Journals & Magazine | IEEE Xplore,"* ieeexplore.ieee.org, April 2022.
[22] J. Leoste, L. Jõgi, T. Õun, L. Pastor, J. San Martín López, and I. Grauberg, "Perceptions about the Future of Integrating Emerging Technologies into Higher Education—The Case of Robotics with Artificial Intelligence," *Computers*, vol. 10, no. 9, p. 110, Sep. 2021, doi: https://doi.org/10.3390/computers10090110
[23] I. Lee and B. Perret, "Preparing High School Teachers to Integrate AI Methods into STEM Classrooms," *Proceedings of the AAAI Conference on Artificial Intelligence*, vol. 36, no. 11, pp. 12783–12791, Jun. 2022, doi: https://doi.org/10.1609/aaai.v36i11.21557
[24] A. Eguchi, "Bringing Robotics in Classrooms," *Robotics in STEM Education*, pp. 3–31, 2017, doi: https://doi.org/10.1007/978-3-319-57786-9_1
[25] X. Chen, G. Cheng, D. Zou, B. Zhong, and H. Xie, "Artificial Intelligent Robots for Precision Education: A Topic Modeling-Based Bibliometric Analysis," *Educational Technology & Society*, vol. 26, no. 1, pp. 171–186, 2023, Available: https://www.jstor.org/stable/48707975
[26] F. Pedro, M. Subosa, A. Rivas, and P. Valverde, "Artificial Intelligence in Education: Challenges and Opportunities for Sustainable Development," *Ministerio De Educación*, 2019, Available: https://repositorio.minedu.gob.pe/handle/20.500.12799/6533
[27] Z. Huang, Y. Shen, J. Li, M. Fey, and C. Brecher, "A Survey on AI-Driven Digital Twins in Industry 4.0: Smart Manufacturing and Advanced Robotics," *Sensors*, vol. 21, no. 19, p. 6340, Sep. 2021, doi: https://doi.org/10.3390/s21196340
[28] M. Kyrarini et al., "A Survey of Robots in Healthcare," *Technologies*, vol. 9, no. 1, p. 8, Jan. 2021, doi: https://doi.org/10.3390/technologies9010008
[29] M. T. Linaza et al., "Data-Driven Artificial Intelligence Applications for Sustainable Precision Agriculture," *Agronomy*, vol. 11, no. 6, p. 1227, Jun. 2021, doi: https://doi.org/10.3390/agronomy11061227
[30] P. J. Singh, and R. De Silva, "Design and Implementation of an Experimental UAV Network | IEEE Conference Publication | IEEE Xplore," *In 2018 international conference on information and communications technology (icoiact)* pp. 168–173, Mar. 2018.
[31] D.-L. Li, L. Zhong, W. Zhu, Z. Xu, Q. Tang, and W. Zhan, "A Survey of Space Robotic Technologies for On-Orbit Assembly," vol. 2022, Sep. 2022, doi: https://doi.org/10.34133/2022/9849170
[32] K. Zhang and A. B. Aslan, "AI Technologies for Education: Recent Research & Future Directions," *Computers and Education: Artificial Intelligence,*

vol. 2, no. 100025, p. 100025, 2021, doi: https://doi.org/10.1016/j.caeai.2021.100025

[33] L. Chen, P. Chen, and Z. Lin, "Artificial Intelligence in Education: A Review," *IEEE Access*, vol. 8, no. 2169–3536, pp. 75264–75278, Apr. 2020, doi: https://doi.org/10.1109/ACCESS.2020.2988510

[34] R. Abduljabbar, H. Dia, S. Liyanage, and S. A. Bagloee, "Applications of Artificial Intelligence in Transport: An Overview," *Sustainability*, vol. 11, no. 1, p. 189, Jan. 2019, doi: https://doi.org/10.3390/su11010189.

[35] S. H. Chung, "Applications of Smart Technologies in Logistics and Transport: A Review," *Transportation Research Part E: Logistics and Transportation Review*, vol. 153, p. 102455, Sep. 2021, doi: https://doi.org/10.1016/j.tre.2021.102455

[36] N. A. Ubina and S.-C. Cheng, "A Review of Unmanned System Technologies with Its Application to Aquaculture Farm Monitoring and Management," *Drones*, vol. 6, no. 1, p. 12, Jan. 2022, doi: https://doi.org/10.3390/drones6010012

[37] P. J. Singh, "*Applications of Deep Learning in Aerial Robotics, Artificial Intelligence*, Chapman and Hall/CRC, pp. 185–209, 2022.

[38] D. Mhlanga, "Industry 4.0 in Finance: The Impact of Artificial Intelligence (AI) on Digital Financial Inclusion," *International Journal of Financial Studies*, vol. 8, no. 3, p. 45, Jul. 2020, doi: https://doi.org/10.3390/ijfs8030045

[39] N. Anantrasirichai and D. Bull, "Artificial Intelligence in the Creative industries: A Review," *Artificial Intelligence Review*, vol. 55, no. 1, Jul. 2021, doi: https://doi.org/10.1007/s10462-021-10039-7

[40] T. Ahmad et al., "Artificial Intelligence in Sustainable Energy Industry: Status Quo, Challenges and Opportunities," *Journal of Cleaner Production*, vol. 289, no. 289, p. 125834, Mar. 2021, doi: https://doi.org/10.1016/j.jclepro.2021.125834

[41] M. Payal, P. Dixit, T. V. M. Sairam, and N. Goyal, "Robotics, AI, and the IoT in Defense Systems," *AI and IoT-Based Intelligent Automation in Robotics*, pp. 109–128, Apr. 2021, doi: https://doi.org/10.1002/9781119711230.ch7

[42] M. Bistron and Z. Piotrowski, "Artificial Intelligence Applications in Military Systems and Their Influence on Sense of Security of Citizens," *Electronics*, vol. 10, no. 7, p. 871, Apr. 2021, doi: https://doi.org/10.3390/electronics10070871

[43] L. Cao, "Artificial Intelligence in Retail: Applications and Value Creation Logics," *International Journal of Retail & Distribution Management*, vol. 49, no. 7, pp. 958–976, 2021.

[44] F. D. Weber, and R. Schütte, "State-of-the-Art and Adoption of Artificial Intelligence in Retailing," *Digital Policy, Regulation and Governance*, vol. 21, no. 3, pp. 264–279, 2019.

[45] V. Dutta and T. Zielińska, "Cybersecurity of Robotic Systems: Leading Challenges and Robotic System Design Methodology," *Electronics*, vol. 10, no. 22, p. 2850, Nov. 2021, doi: https://doi.org/10.3390/electronics10222850

[46] J. P. A. Yaacoub, H. N. Noura, O. Salman, and A. Chehab, "Robotics Cyber Security: Vulnerabilities, Attacks, Countermeasures, and Recommendations," *International Journal of Information Security*, vol. 21, no. 21, Mar. 2021, doi: https://doi.org/10.1007/s10207-021-00545-8

Chapter 9

Applications of robots in assistive technology

Yamiko Daniel Banda, Tuhina Raj, and Rashmi Priyadarshini

9.1 INTRODUCTION

The rise in robotics has transformed industries, including assistive technology. In years, many countries are expected to see growth in their elderly populations. Older individuals often have healthcare needs. Many elders who live unaccompanied make it difficult to speedily notify others in situations of emergencies like falls or in case of accidents. This solitary life also increases the probability of depression and continuing loneliness, which has been associated with health issues connected with obesity or smoking. These robots can engage users in communicating activities, offering reminders for medication or appointments and even entertainment and emotional support.

Assistive technology involves a wide range of devices, services as well as systems that are needed to support and enhance the independence and comfort of people with disabilities and chronic conditions. Robots can help in transforming numerous traditional assistive devices by enhancing abilities such as semi-autonomous navigation, environmental sensors, and adaptive motor control. Robots have the potential to support people with disabilities or age-related limits in performing everyday responsibilities helping them to develop their quality of life. From robotic exoskeletons that support individuals with mobility impairments to robotic arms that aid with feeding and personal care responsibilities, the applications in these areas are increasing.

One of the main areas of focus is evolving robotic assistive technologies for movement and physical tasks. Powered exoskeletons and robotic prosthetic limbs can be combined with biosensors and algorithms to enable natural and innovative control structures. Such a device purpose is to bring back mobility and enhance the operating abilities to individuals who lack full limb functions. Robots are also offering physical support without being worn by the user. Robotic walkers, intelligent canes and autonomous wheelchairs can navigate environments, evade obstacles, and move over compact spaces.

Robots are similarly being merged into assistive technologies for the home. Intelligent robotic arms can be mounted as helpers to pick up to

DOI: 10.1201/9781003537083-9

operate objects for users with partial upper body function. Mobile robots can automatically fetch demanded items, perform household works, and deliver monitoring and emergency responses. Such technologies aim to permit users to monitor and control their home environments. For those with sensory impairments such as vision or hearing loss, robotic technologies are allowing assistive devices with added intelligent processing and situational awareness skills. Robots can deliver object recognition and navigation help for the blind. Robotic hearing assistants can employ speech recognition and natural language processing to answer audio commands and can engage in cooperative communication.

Assistive robotics is also making advancement in intellectual and social support applications. Robots can offer therapy, coaching, and educational provision to individuals with brainy disabilities. Their abilities for natural communication and human-like interaction can afford engaging support for building life skills. Social assistive robots can support operators build confidence, initiate conversations, and increase their social connections.

More research is needed to improve the robotic capabilities of user interaction systems and integration with artificial intelligence computing. Seamless incorporation into everyday routines remains a challenge. However, the possibilities are promising for robotic technologies to profoundly transform assistive tools and empower individuals with disabilities to live more self-reliantly, engage socially, and fully participate in their communities. The future of assistive technologies will increasingly integrate intelligent robotic functions to provide users with enhanced capabilities tailored to their unique needs and preferences.

In the below sections, this study will support various applications that already exist in the commercial market in today's era.

9.2 TOYOTA HUMAN SUPPORT ROBOT

Toyota Motor Corporation has been developing assistive robots aimed at providing support services for the elderly and people with disabilities. A major focus of their research has been on creating robots that can physically assist users with simple everyday tasks. This aligns with Toyota's goal of mobilizing robotics' technology to promote independent living and improve quality of life. One of Toyota's key initiatives in this area is the Human Support Robot (HSR).

Unveiled in 2017, the HSR is a small one-armed robot mounted on a wheeled base. It incorporates advanced sensor systems, artificial intelligence, and Toyota's extensive automation and manufacturing expertise. The HSR can autonomously navigate environments using its sensor array including cameras and LiDAR. This allows it to move safely, avoid obstacles, and travel alongside humans. The robot's single arm has a gripper designed to grasp and manipulate a variety of common household objects. The gripper can pick up items from the floor or shelves and bring them to the user. This assists users who have difficulty bending or reaching. The arm has compliant joints and forces control sensors to handle objects gently and safely to pick up and retrieve everything from dropped phones to food from the refrigerator. The interface consists of voice control, facial expressions, and an interactive touchscreen. This allows the user to give the robot commands and feedback. Figure 9.1 depicts the HSR from Toyota.

The built-in AI computing provides object and speech recognition capabilities aiming to understand natural language commands and engage in human-like two-way communication. With its friendly expressions and voice, Toyota strives to make the robot's assistance seem warm and welcoming to physical tasks and perform monitoring and communication functions. Its sensors can gather environmental data and transmit it to the user or caregivers. It can also serve as a mobile interface for video calls and home control networks [1]. As an embodied assistance robot, the HSR can

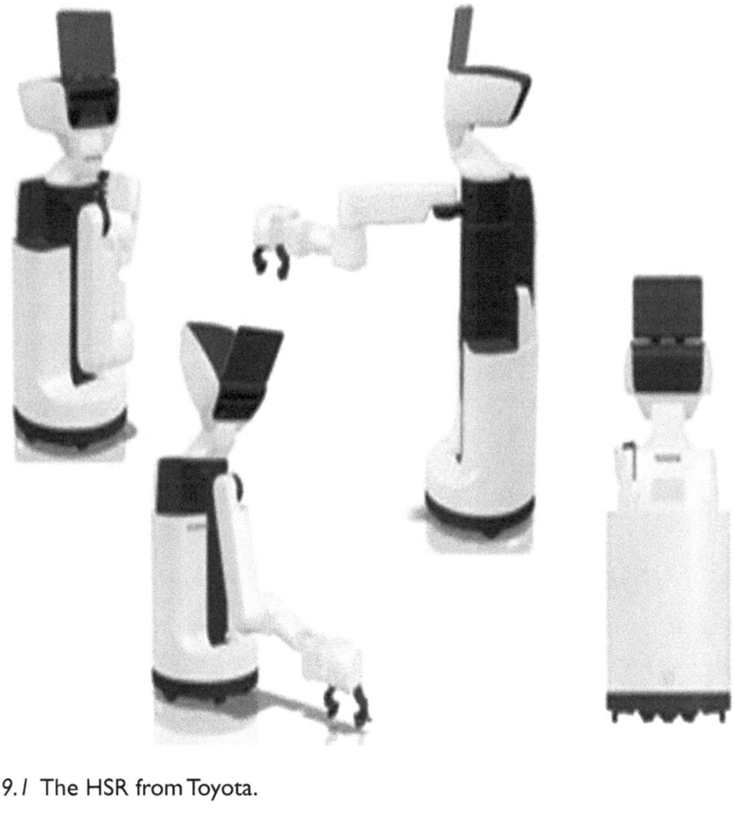

Figure 9.1 The HSR from Toyota.

move to various locations in the home and act as a helper that is nearby throughout the day. Toyota views the HSR as a general-purpose platform that could take on a wide range of assistive roles with capabilities that could potentially be tailored to users with diverse mobility or cognitive needs.

Toyota is exploring integrating the robot into smart home infrastructure. This would allow it to collaborate with other connected devices and services including AI virtual assistants. The capabilities and flexible AI platform of the robot could lead to valuable assistive applications. As a mobile physical assistant, the HSR demonstrates the potential for service robots to take on helpful roles in home environments. Toyota intends to continue refining the HSR's capabilities and testing real-world implementations to turn this technology into practical solutions [2].

9.3 XANDERGLASSES

Xander are innovative assistive technology designed to help individuals with severe hearing loss to understand speech more effectively and has a collaboration in place with the Veterans Health Administration, enabling us to work closely with physicians and staff at VA Medical Centers to offer XanderGlasses to qualified Veterans with hearing challenges. XanderGlasses utilizes expanded reality (AR) innovation to work on face-to-face correspondence for individuals impacted by hearing misfortune. They likewise can decipher discourse into text and show subtitles progressively, furnishing wearers with a consistent method for following discussions without the need to peer down at any extra gadgets. The glasses are fueled by Vuzix Safeguard savvy glasses, which are known for their lightweight and agreeable plan that feels normal to wear. XanderGlasses can also be used in various settings, including at home, work or in any noisy public venues, to help the individuals with hearing loss to understand who is speaking to them. By providing real-time automatic captions of conversations, these glasses aim to increase confidence and comfort during social communications, which enhances communication involvement for users. The glasses have also gathered media coverage and positive testimonials from operators, which could highlight their potential to advance the lives of individuals with hearing challenges [3]. Figure 9.2 shows the XanderGlasses which are presently in the marketplace.

9.4 NAQI NEURAL EARBUDS BY NAQI LOGIX

Naqi Neural Earbuds are a type of earbud established by a company called Naqi Logix and are planned to deliver a unique solution in the field of assistive technology. Naqi Neural Earbuds by Naqi Logix is an advanced invention in assistive technology that transforms human–machine interaction.

These earbuds present innovative and advanced features that have not been seen before in assistive technology, providing an extra continuous and

Figure 9.2 XanderGlasses.

efficient way for users to cooperate with machines. Naqi Neural Earbuds are a kind of earbud, which allows users to control and interact with several digital devices, smart home IoT devices, robots and wheelchairs to offer a hands-free as well as voice-free, and screen-free platform, which means users can control these devices without using their voice, touch, screens, or cameras. The technology creates a new way for users to cooperate with technology, allowing for a more seamless and intuitive experience. Neural earbuds have been applauded for their capability to control nearly anything digital, making them appropriate for a wide range of operators as well as the disabled, elderly, gamers, e-sports professionals, and first responders. Figure 9.3 represents the model design for the earbuds.

The technology behind Naqi Neural Earbuds uses advanced technology that can sense brain waves, muscle impulses, and other signals from the user's body. These signals are then changed into commands that can control devices such as wheelchairs, drones, computers, and AR/VR systems. The impact of Naqi Neural Earbuds is a front-line solution that allows users to interact with technology flawlessly and naturally with a wide impact across various domains, including accessibility, gaming, and robotics' control, and deliver a hands-free interaction method that is key for individuals who cannot use traditional input methods, such as those with disabilities or the

Figure 9.3 Naqi Neural Earbuds by Naqi Logix.

aged. While not yet commercially available, Naqi earbuds have attracted major attention. They were also included in TIME Magazine's "200 Best Inventions of 2023" and showcased at CES 2024 [4].

9.5 THE CP-WALKER 2.0

Cerebral palsy is the most common neuromuscular disease, affecting 2–3 out of every 1000 live births worldwide. It is caused by the damage to the developing brain often before or during birth, leading to impaired muscle coordination as well as posture control. The new CP-Walker integrates the inventions like a helmet that autogenerates the answer to the wearer's brain signals providing personalized support as they walk [5]. This assistance provides more upright, stable, and balanced movements compared to traditional walkers. The possible scope of assistances of the CP-Walker 2.0 and similar robotic exoskeletons are playing major roles in changing the advancements of assistive technology. Not only can they directly assist mobility, but the synchronized movement patterns they support may benefit and strengthen muscles and neural paths. This could lead to better posture and walking ability even without using any of the device. The upright stance can also enable cognitive and social assistance when compared to using wheelchairs. Exoskeleton-assisted walking may improve circulation and bone density as well as a leading to drop in secondary complications [6].

Robotic exoskeletons are one module of a wide-ranging cerebral palsy treatment plan. Other therapies like physical therapy, occupational therapy, and medicines are critical too. Future technological advances may allow exoskeletons to provide even more tailored, adaptable support by interfacing with the wearer's brain signals. Already, EEG headsets can interpret some movement intentions. As brain–computer interfaces improve, exoskeletons can respond to thoughts in real-time.

Stem cell research is another exciting area that may someday benefit cerebral palsy patients. Studies injecting bone marrow stem cells intrathecally have demonstrated neurological improvements in children with cerebral palsy. The stem cells are thought to migrate to sites of brain injury and differentiate into neurons or supporting cells. This may repair damage and forge new neural connections. While still experimental, it hints at the vast potential of stem cells to treat developmental brain disorders. More research is underway, often in combination with physical rehabilitation [7]. In the future, cerebral palsy treatment will likely involve an integrative approach including robotic assistance, stem cell therapy, physical therapy, and more. The recent progress in robotic exoskeletons is encouraging and reflects broader advances in rehabilitation technology. Figure 9.4 represents the CP-Walker 2.0 and how it is assisting mobility.

9.6 REHABILITATION EXOSKELETONS FOR THE LOWER LIMBS

Lower limb exoskeletons are emerging as an innovative technology for assisting rehabilitation for individuals with mobility impairments or weakness affecting their legs and feet. These robotic devices are worn by the user and integrate with the body's movements to provide stability, strength augmentation, or guided mobility therapy. There is a wide range of potential applications for lower limb rehabilitation exoskeletons that can transform traditional physical therapy practices. For patients re-learning to walk after injuries, strokes, or other neurological conditions, exoskeletons provide essential physical support and assistance with weight-bearing and balance. Devices like the Ekso GT allow patients to practice walking using crutches for stability while the exoskeleton takes on some of the bodyweight load. Such devices aim to rebuild strength, coordination, and confidence during gait training. Therapists can control and customize levels of assistance as patients progress. This intensive training method has shown excellent

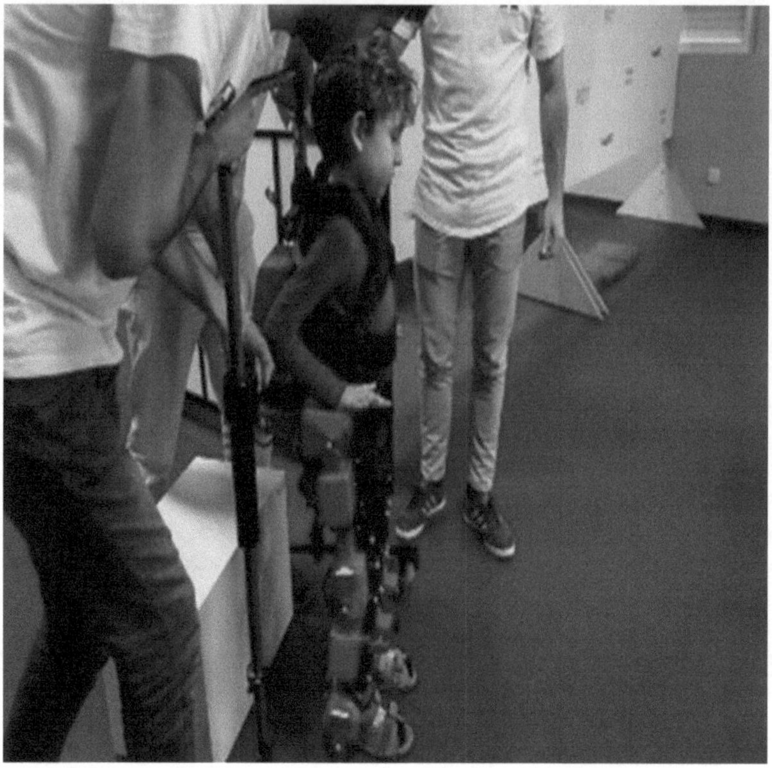

Figure 9.4 CP-Walker 2.0.

outcomes in restoring functional mobility [8]. Another key application is using exoskeletons to assist specific maneuvers that are difficult for patients re-learning to walk. For example, systems like the Tibion device can assist with sit-to-stand transitions and stair-climbing activities. The joint actuators and sensors detect the user's movements and provide an extra boost at key phases. This allows patients to practice proper techniques and build lower limb strength. Real-time assistance helps prevent falls and injuries during the rehabilitation process [9].

For mobility-impaired individuals unable to walk independently, exoskeletons can provide basic upright mobility. Wheelchair-bound patients value the ability to stand and move short distances in an upright position for physical and psychosocial benefits. Standing devices like the ReWalk restore a degree of mobility while providing cardiovascular and bone health benefits compared to sitting long-term. Some systems allow supported standing transitions and basic walking on level surfaces.

Powered orthotic devices are useful to assist detailed gait abnormalities in patients with stroke, cerebral palsy, multiple sclerosis, and other conditions. For example, systems like the MyPro delivers powered support at the knee joint to compensate for foot drop during the swing phase of gait. Other orthoses can help calm abnormal gaits and transmit limb movements into more normal patterns through directed force feedback. Patients get the experience of improved gait even though rebuilding their muscle control. Finally, exoskeletons are being used in the therapy of elderly patients with overall muscular weakness and frailty affecting movement [10]. Devices with active power support at many joints can provide support during physical activity exercises to recover strength, balance, and function. This allows tailoring resistance levels to each patient's capabilities so they can complete movements and exercise regimens not otherwise possible. By enabling more successful movement experiences, exoskeletons can help seniors partially regain mobility and reduce fall risks. Advanced robotic lower limb exoskeletons are transforming rehabilitation by enabling new methods of gait training and assisted mobility. Their applications span from short-term restorative therapy to daily living assistance for the chronically mobility impaired. Continued development aims to make such devices more wearable, intuitive, and compatible with users' emotions and needs [11]. Figure 9.5 represents the plan analysis for the elderly and people with disability.

9.7 THE REWALK PERSONAL 6.0 EXOSKELETON

ReWalk Robotics is a wearable robotic exoskeleton designed to provide powered hip and knee motion, enabling individuals with spinal cord injuries (SCI) to stand upright, walk, turn, and even climb and descend stairs. The exoskeleton is programmed to mimic a natural gait pattern, enabling users to walk in a way that closely resembles how they would walk without the device. This functionality of the ReWalk system promotes independent and controlled walking, giving individuals with SCI the ability to move around more freely and perform daily activities. The device is shaped exclusively for each user to ensure top safety, function, and joint alignment allows for an adapted fit, and ensures that the exoskeleton delivers the best possible provision and assistance to the user. The exoskeleton is designed to be worn all day, both at home and in the community, providing nonstop support and movement for people with SCI [12].

These benefits consist of a reduced requirement for certain medications, improved bowel and bladder function, enhanced mental health, better sleep quality, reduced tiredness, decreased body fat, pain relief, better posture, better balance, and overall well-being and quality of life for individuals with SCI by addressing several aspects of their health and flexibility. Users of the ReWalk system have stated important psychological assistance from using the exoskeleton as being able to stand and walk again with the support of

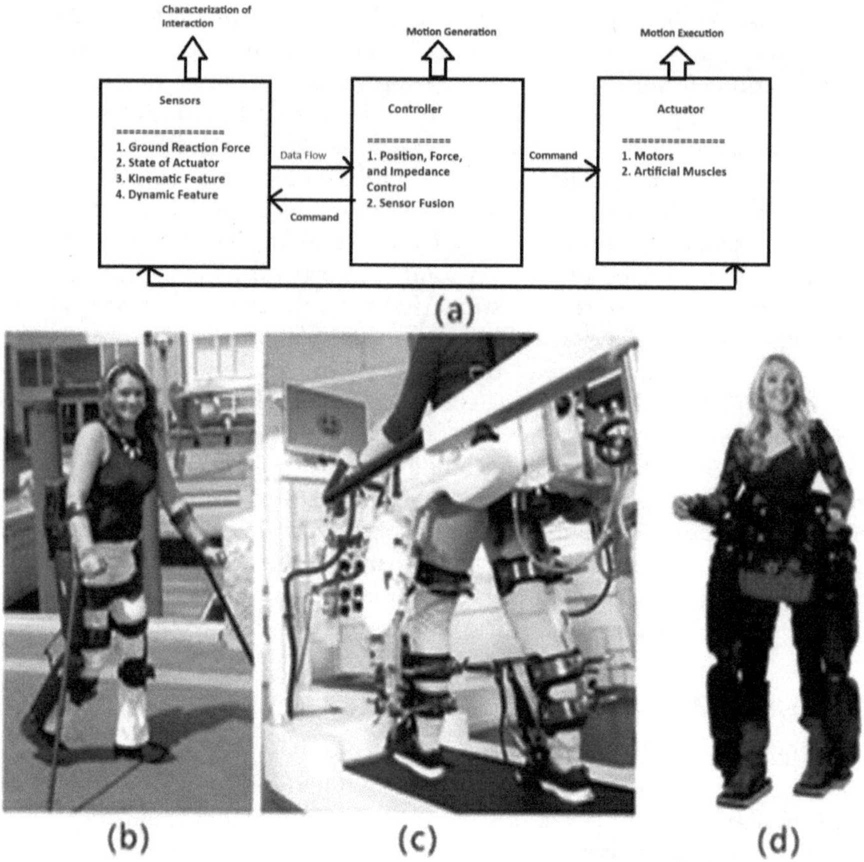

Figure 9.5 Rehabilitation exoskeletons for the lower limbs.

the device has brought them joy and a sense of achievement with the ability to regain movement and individuality can have a positive impact on an individual's mental well-being, boosting their confidence and overall emotional state. On maintenance and support, the ReWalk system assists users in seeking reimbursement for personal systems, helping them navigate the process of obtaining financial exposure for the device [13].

In addition, the system also offers a full maintenance and warranty program after purchase, which warrants that the user keeps on receiving continuous support and assistance with any technical issues or maintenance that could arise in the future. The ReWalk Personal 6.0 Exoskeleton displays an advanced technology that aims to enhance the mobility and quality of life for individuals in cases of people with spinal cord injuries. Figure 9.6 shows the ReWalk Personal 6.0 Exoskeleton and how it could be worn off.

Figure 9.6 The ReWalk Personal 6.0 Exoskeleton.

9.8 UTAH BIONIC LEG

Assistant Professor Tommaso Lenzi and his colleagues have developed a prosthetic limb that replicates the motion of normal leg and knee functions, which could help in the simpler movement for users, particularly on stairs. The prosthetic limb is engineered to precisely imitate the movements of a normal leg and knee, enabling a more natural and seamless motion, and facilitating simpler navigation of stairs and difficult terrains. The prosthetic leg resemble the movement of a normal leg to improve comfort and efficiency for amputees while walking [14]. Assistant Professor Tommaso Lenzi and his colleagues have created a prosthetic leg that mimics the motion of a normal leg and knee, facilitating mobility for users, particularly while climbing stairs. Because of the prosthetic leg's strong resemblance to the original knee and leg, users may walk more naturally and fluidly, which makes it simpler for them to negotiate stairs and other difficult terrain. The prosthetic leg attempts to provide amputees a more efficient and comfortable gait by imitating the movement of a normal leg [15].

9.9 MOTION CONTROL ILIMB ULTRA

The iLimb Ultra is a type of prosthetic hand produced by a company called Touch Bionics designed to provide a high level of dexterity and control to individuals who have lost their natural hand or fingers. The hand has many fingers that can move and grip items, enabling the user to engage with their surroundings more naturally and effectively. The device features many autonomously powered digits that may operate separately, enabling a diverse array of gripping and handling operations. Users can perform tasks that demand precise motor skills, like picking up small objects or using tools with multiple fingers that mimic natural finger movements such as flexion, extension, and opposition. The user can utilize the prosthetic hand to execute various grasping activities, including pinching, gripping, or holding items.

The fingers of the prosthetic hand are designed to be able to exert a certain amount of force, allowing the user to securely hold objects of different sizes and weights to perform everyday tasks, such as picking up a cup, typing on a keyboard, or even playing a musical instrument. Electronically rotatable thumb, with manual override, automatically switches between lateral and oppositional positions with 18 different automated grip options available in four sizes – extra small, small, medium, and large. Extra small digits are aluminum for lightweight durability with muscle control it uses specific muscle signals called triggers to instruct the hand to activate a specific grip [16]. Figure 9.7 showcases the model of motion control iLimb Ultra assistive technology.

9.10 TATUM T1

The recently made Tatum T1 robot is a critical improvement in assistive innovation for the people who are both hard of hearing and visually impaired. Tatum Mechanical technology fostered an extremely coordinated automated arm that empowers hard of hearing visually impaired people to grasp discussions through material American Gesture based communication (ASL) translations. By taking advantage of the arm and detecting its exact movements, clients might decipher messages, books, news, and sites. This grants constant correspondence and data access without requiring a human translation. Significant advancement behind the Tatum T1 is its strong ASL interpretation innovation. This restrictive program can make an interpretation of composed English text into precise, syntactically amazing ASL. Catching the nuances of ASL sentence structure and grammar rules was a significant mechanical test requiring cautious assessment of language frameworks.

Figure 9.7 Motion Control iLimb Ultra.

The outcome of this calculation is a worthy representative for the capacity of Tatum's specialized group. It presents the potential for future applications also, including programmed verbal interpretation of material ASL. For hard-of-hearing visually impaired clients, the advantages of the Tatum T1 are colossal. It empowers independence and protection for individual discussions without relying upon a mediator. The ability to uninhibitedly access news and data likewise creates autonomy and social consideration. These characteristics can extensively work on personal satisfaction and decrease depression for the hard of hearing visually impaired individuals [17]. Moreover, the material ASL translations work on a consciousness of language designs and phrasing. This can maybe help support lingering hearing or vision.

Looking forward, Tatum desires to foster the T1's material input for much more delicate and regular correspondence. Sensors could permit movable grasp strength and movement speed to acclimate to every client. High-level tension criticism through the arm's silicone skin could more readily address finger composing and profound nuances. Tatum likewise tries to upgrade the T1's capacity to incorporate a greater jargon and language intricacy to deal with extra use cases. In the meantime, different organizations are creating practically identical innovations, such as haptic gloves making communication via gestures and movements in the palm. As assistive mechanical innovation keeps propelling, it might sometimes permit hard of hearing visually impaired people to draw in with the climate similar to located and hearing people. Incorporating versatile, visual acknowledgment and discourse handling could offer ongoing natural data utilizing ASL. At some point, material deciphering robots like the Tatum T1 might become far-reaching availability helps to permit hard of hearing visually impaired people to investigate, blend, and concentrate as openly as anyone. At present, the Tatum T1 stays in model stages requiring a greater turn of events and testing before broad conveyance. Yet, it signals significant advancement toward a hopeful goal – changing correspondence for the roughly 40,000 hard of hearing and visually impaired Americans to give them opportunity, contribution, and uniformity. Driven by ongoing improvements in man-made reasoning, AI, and haptic innovation, the T1 clues to an impending future where limits need not limit points of view or goals [18]. Figure 9.8 addresses the model of the Tatum T1.

9.11 THE AUGMENTED CANE

The recently made Augmented Cane from Stanford College denotes a critical jump in assistive innovation for the sight impaired. Not at all like common white sticks, this brilliant stick integrates sensors and route advancements to give better ecological mindfulness and development support. As per research distributed in science advanced mechanics, this original innovation could adjust how the visually impaired and outwardly tested travel and collaborate with their general surroundings. The Increased Stick's significant highlights are cameras, LiDAR sensors, GPS, and an omnidirectional wheel. LiDAR utilizes reflected laser heartbeats to plan the area of encompassing items. This information assists the stick with perceiving hindrances and moving around them. The wheel latently lines up with the

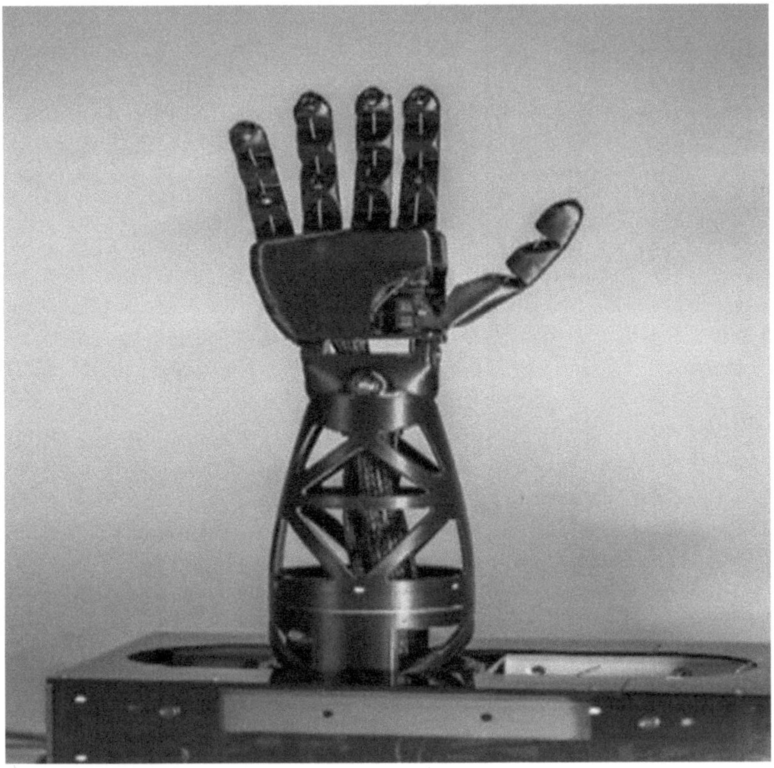

Figure 9.8 Tatum T1.

client's strolling direction, yet may likewise convey haptic criticism bumps to help route. These characteristics permit outwardly hindered people to beat obstructions and stay away from gambles that a standard white stick could miss. For clients, this suggests expanded versatility and autonomy. They can cross obscure environmental elements and testing conditions serenely and safely [19].

The Augmented Cane further develops investigation and diminishes reliance on guides. It is lightweight at simply 1.4 lbs, making it simple to travel and make due. The innovative upgrade additionally does not impede the stick's regular material information. Clients can in any case follow walls, sense walkway limits, and identify surface surfaces and geological contrasts. Contrasted with before brilliant stick drives, the Increased Stick is intended to join convenience, straightforwardness, and cost. Its $400 value makes it available to many. The makers' open-source dispersion of schematics and code upholds progressing headways. Future forms might consolidate new sensors, give customized settings, and empower local area remarks.

Looking ahead, the technique might be advanced even more with autonomous navigation in difficult settings. AI systems might map the best routes on the fly while avoiding impediments. Integration with smart home and city infrastructure might allow the cane to regulate lights, unlock doors, and integrate with public transit and ridesharing. Despite its tremendous promise, considerable training and adaption will be necessary for general adoption. Visually challenged users have depended on plain white canes for decades.

Transitioning to high-tech navigation aids will entail acquiring new skills and gaining trust. Hands-on training programs and community support can smooth this process. But the advantages will make the work worthwhile. The Augmented Cane exemplifies how technology can be leveraged to expand abilities and transform lives. While still in the early stages, its development underscores the vast potential to apply robotics and AI to reduce mobility barriers. As rapid advancements continue in sensors, machine learning, and haptic feedback, the future looks brighter for innovative assistive tools benefiting the visually impaired. The Augmented Cane is just one pioneering step toward that future of inclusion and independence. Figure 9.9 demonstrates the utility of augmented canes [20].

9.12 OPTIMUS 2/OPTIMUS 2 RS

The revamped Optimus 2 electric wheelchair offers powerful off-road capability with stronger batteries and a top speed of up to 15 km/h. Software adjustments have enhanced not only the extended range of up to 111 km in the 6 km/h version but also improved driving stability and a high-intensity LED lighting system ensures better visibility on cloudy days. Invacare Corporation in the medical equipment industry, specializes in the development of smart wheelchairs that incorporate wireless and sensor technologies to enhance mobility for individuals with limited mobility [21].

The market for smart wheelchairs is expected to grow due to factors such as the increasing demand for automated medical devices, the rise in the global geriatric population, and the utilization of smart wheelchairs in various sectors like airports, hospitals, and educational institutions. Additionally, advancements in technology, such as artificial intelligence and machine learning, offer growth opportunities in the smart wheelchair market. Advancements in technology, including artificial intelligence and machine learning, have paved the way for intelligent-based systems in smart wheelchairs offering growth prospects for participants in the smart wheelchair market by attracting more consumers who appreciate smart devices and wearable technology.

The increasing number of senior citizens and individuals with physical disabilities has driven the development of intelligent-based systems in smart wheelchairs to cater to a growing population with limited mobility. Among

Figure 9.9 AR Cane.

these people, those who have problems with mobility, rely on wheelchairs to assist their movement. Manual wheelchairs are usually propelled either by the user or an attendant. If the user wants to operate it by themselves, they should have good upper body function and stamina [22].

Manual wheelchairs are lighter in weight, easier to transport without special equipment, and require less maintenance than power wheelchairs or scooters. Power wheelchairs sometimes included or referred to as scooters are indicated for those who cannot manually propel a wheelchair and who have the cognitive ability and coordination needed to safely propel a power device. They may be appropriate for patients with upper limb weakness, cardiopulmonary disease, poor trunk stability, upper limb pain, limited endurance, or obesity. These innovations aim to enhance accessibility and inclusion for individuals with disabilities while promoting social and economic integration. Figure 9.10 shows the prototype for the Optimus 2 electric wheelchairs [23].

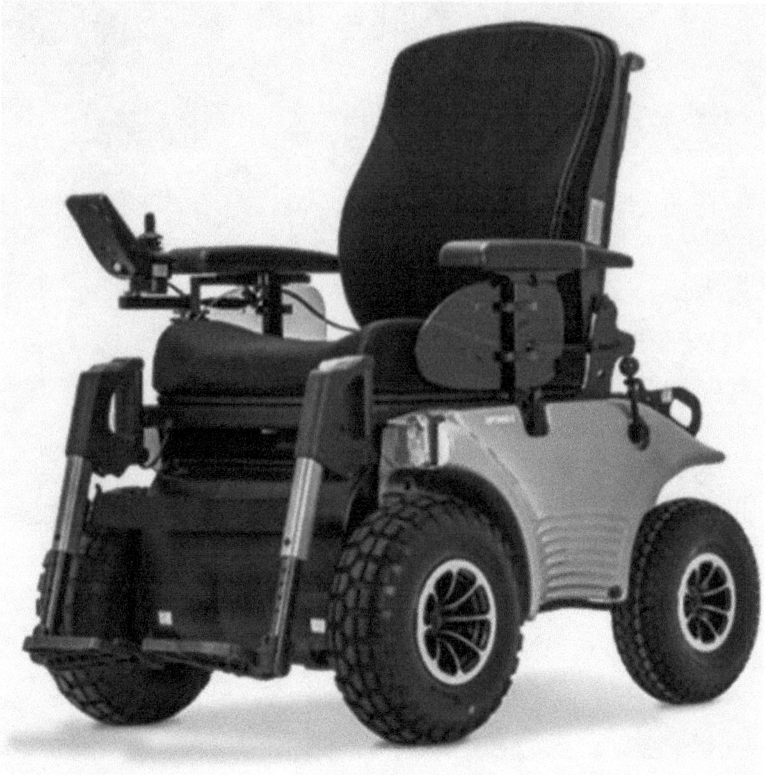

Figure 9.10 Optimus 2.

9.13 AUTONOMOUS VEHICLES

An autonomous car is a vehicle capable of sensing its environment and operating without human involvement. A human passenger is not required to take control of the vehicle at any time, nor is a human passenger required to be present in the vehicle at all and can go anywhere a traditional car goes and do everything that an experienced human driver does. Autonomous cars rely on sensors, actuators, complex algorithms, machine learning systems, and powerful processors to execute software [24].

Autonomous vehicles generate and update a map of their environment using a range of sensors placed in various locations on the vehicle. Radar sensors track the location of close cars. Video cameras identify traffic signals, interpret road signs, monitor other cars, and search for pedestrians. LiDAR sensors use light pulses to calculate distances, detect road boundaries, and recognize lane markers in a car's surroundings. Ultrasonic sensors located in the wheels identify obstacles and other cars when parking. The Society of

Automotive Engineers (SAE) has developed a system to classify the various levels of vehicle autonomy, which consist of six categories ranging from no automation to complete automation. These levels are used to classify the capabilities and limitations of autonomous vehicles. Level 0 represents vehicles that have no automated driving features. However, these vehicles may still have driver assistance features, which are technologies that provide support to the driver but do not take full control of the vehicle and the driver is responsible for all driving tasks and must perform them without any automated assistance. Level 1 represents vehicles with advanced driver assistance systems (ADAS).

ADAS technologies provide support to the driver during specific driving tasks, such as acceleration, braking, and steering. However, the driver is still actively involved and must remain attentive while using these systems, the driver is ultimately responsible for the safe operation of the vehicle. Level 2 represents vehicles that have partial driving automation capabilities where vehicles can control steering, acceleration, and braking functions.

However, the driver must remain attentive and be ready to take control of the vehicle at any time. Level 2 automation requires the driver to constantly monitor the driving environment and be prepared to intervene if necessary. Level 3 represents vehicles that can perform driving tasks under certain conditions and are capable of controlling all aspects of driving, including steering, acceleration, and braking, in specific situations.

However, the driver must be prepared to take control of the vehicle when the system requests it which allows the driver to disengage from actively driving the vehicle but still requires them to be ready to intervene if needed. Level 4 represents vehicles that can operate autonomously under certain conditions as they can perform all driving tasks without human intervention in specific environments or situations. Level 4 automation provides a high level of autonomy but still requires the driver to be available as a backup, if the system encounters a situation it cannot handle. Level 5 represents vehicles that can perform any driving function a human driver can, without requiring human intervention as these vehicles are fully autonomous and do not have a steering wheel or pedals. Level 5 automation allows passengers to completely disengage from driving tasks and rely on the vehicle to handle all aspects of driving. The level of automation is the highest level of autonomy and represents a fully self-driving vehicle [25, 26].

9.14 VULCAN INTELLIGENT WHEELCHAIRS

The University of Michigan's Vulcan project focuses on developing intelligent wheelchairs as assistive technology to improve mobility and independence for people with disabilities. Named after the Roman god Vulcan, who was lame, the goal is to create smart wheelchairs that can navigate environments and assist users more effectively. To achieve this vision, the robot must have rich knowledge of the task domain and the knowledge must be used to enable fluid communication between humans and machines. Navigation is one domain where the framework of knowledge is well enough known for a robot to be a good companion in addressing actual requirements [27].

The venture involves two kinds of navigational robots: an insightful mechanical wheelchair and a telepresence robot. The wise wheelchair has been created to pass its client safely on to indicated areas. It builds a mental guide to get a handle on the plan of its current circumstance and review ways, permitting it to move about really. The telepresence robot moreover creates mental guides but, however, passes its impacts on to a far-off human administrator rather than an installed client. The two kinds of robots gain from venture out encounters to improve their navigational capacity after some time. Viable correspondence is critical for reinforcing connections among clients and assistive innovation. The robots are intended to talk with individuals in a characteristic way that advance powerful conversation in regards to route objectives, ecological conditions, and expected obstructions. This permits the robots to grasp human needs and convey proper assistance. The correspondence capacities try to furnish clients with a superior feeling of opportunity and control and may be used for exercises like conveying supplies in emergency clinics where the seats' capacity to arrange bound puts freely would be worthwhile. Telepresence robots can give significant assistance in conditions when it is hazardous or challenging for individuals to be actually present [28].

For instance, they could give crisis staff improved situational mindfulness in catastrophe zones without forfeiting lives. The Vulcan project shows the colossal capability of advanced mechanics to change assistive innovations. Brilliant wheelchairs can altogether upgrade versatility and freedom for the handicapped. Further developed correspondence connection points will extend communications among people and machines, permitting robots to give help that is more receptive to individual requirements. Telepresence and far off presence robots will empower individuals to, for all intents and purposes, project themselves into settings where they cannot be truly present, significantly extending their scope and abilities. Research in assistive mechanical technology likewise creates advancements that spread to different fields. For example, the navigational calculations being created could be applied to independent vehicles. Discourse acknowledgment and regular language handling fueling correspondence interfaces have wide applications. Working on mechanical versatility and control helps the more extensive objective of building able and valuable broadly useful robots. The Vulcan undertaking and others like it are growing the skylines of assistive innovation while additionally pushing advanced mechanics forward [29]. Figure 9.11 addresses the plan for Vulcan Smart wheelchairs.

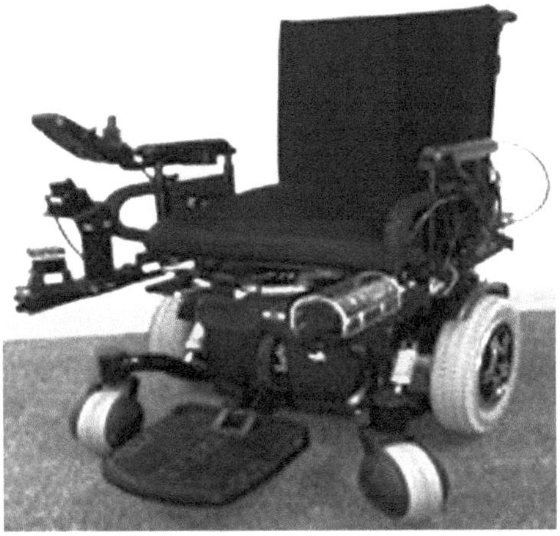

Figure 9.11 Vulcan 1.0 was the Intelligent Wheelchair at the University of Texas at Austin and Vulcan 2.0 is the Intelligent Wheelchair at the University of Michigan.

9.15 THE CO-OPERATIVE LOCOMOTION AIDE (COOL AIDE)

COOL Aide is a three-wheeled passive smart walker equipped with mapping and obstacle detection systems, navigation and guidance algorithms, force sensors on its handlebars, and a Laser Range Finder (LRF) to estimate the user's desired direction. Despite being a passive walker, it incorporates shared control strategies by allowing control to be shared between the platform and the user. This shared control system involves navigation and guidance algorithms in conjunction with shared control systems, enhancing the user's interaction with the walker. COOL Aide utilizes shared control techniques meaning that the steering control is shared between the user and the walker [30]. These techniques are based on aspects including biomechanical and spatiotemporal characteristics, user goals of movement, and environmental restrictions. By combining shared control systems, the walker seeks to provide effective support, comfort, and safety throughout rehabilitation and assistance duties. It also offers numerous control schemes to suit the individual demands of the user. It primarily focuses on boosting the walking experience for the elderly, which does so through advanced features including mapping, obstacle detection, and navigation help. Additionally, it integrates shared control mechanisms to give effective help, comfort, and safety throughout rehabilitation and assistance duties [27].

9.16 AGORA WALKER

The AGoRA Walker is a high-level astute automated walker created to give versatility help and backing to those with step debilitations or trouble strolling freely. Outfitted with strong sensors, planning abilities, and calculations, this brilliant walker looks to increment client portability, security, and independence. At the center of the AGoRA Walker is its detecting and planning advances. It utilizes LiDAR and different sensors to sweep and guide the encompassing region in 3D ceaselessly. It empowers the walker to foster a full spatial guide of pathways, hindrances, things, and that is just the beginning. With constant updates, the guide empowers independent routes by ascertaining amazing courses and staying away from risks. The walker's sensors likewise offer significant information on the client's movements and powers on the handlebars, permitting better control. The AGoRA Walker's route innovation consolidates sensor information with artificial intelligence calculations to lead the client along wanted courses consistently and reliably.

The PCs perceive the client's versatility purposes in light of power examples and body signals. Coordinating this with the natural guide, the walker may tenderly alter direction on a case-by-case basis while as yet empowering the client to keep undeniable level control. This coordinated navigational control among people and innovation looks to advance security and certainty while strolling. To help regular client associations, the Public square Walker offers a natural human–robot interface. This integrates touchscreen screens, speakers, vibration signs, and sensors that make the walker receptive to voice requests, motions, and different information sources. The multimodal interface attempts to recreate a human strolling friend that is delicate to the client's requests. All collaborations are intended to be simple and require little client exertion for availability. Broad testing with sound workers has allowed improvement of the walker's control plans and connection points for best client solace, utilization, and delight. The common navigational control specifically permits clients critical productivity in driving the walker while feeling in charge of their development [31]. Generally, the Public square Walker plans to give actual help, yet additionally more noteworthy independence and fearlessness while strolling.

With its insightful helping abilities, the AGoRA Walker might assist people with confined portability and equilibrium concerns cross regions and execute day to day strolling errands all the more safely and without any problem. For senior grown-ups at risk of falls, the walker's directing and backing can assist them with safeguarding opportunity in development longer. For patients going through actual restoration, the walker can flawlessly help with strolling activities and exercises to supplement treatment. The Public square Walker's capacities make it profoundly flexible across populaces with gentle to direct versatility disabilities. Past assistive applications, the Public square Walker additionally exhibits state-of-the-art human–robot

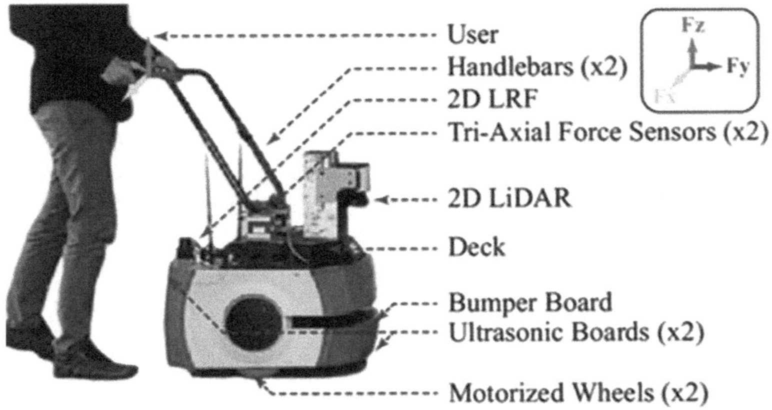

Figure 9.12 AGoRA Walker.

association and route capacities that could be applied to other mechanical gadgets. The points of interaction and calculations can assist with acknowledging more natural, client-driven plans across administration mechanical technology, medical care robots, and, surprisingly, independent vehicles. As an exploration stage, the Public square Walker features promising bearings in making arising advances more receptive to and cooperative with human clients and their requirements [32] (see Figure 9.12).

9.17 TESLA ADDED CUSTOM DRIVER PROFILES IN 2022

Tesla, Inc. is a pioneering American company founded in 2003 by engineers Martin Eberhard and Marc Tarpenning, with a vision to demonstrate that electric vehicles could outperform traditional gasoline cars in terms of speed, fun, and efficiency. In 2022, Tesla introduced custom driver profiles in their vehicles, such as the Model 3 and Model X, to personalize settings and preferences for different drivers. These driver profiles allow for the customization of various aspects of the vehicle, including the driver's seat, steering wheel position, mirrors, and other preferences to enhance the driving experience by automatically adjusting these settings based on individual profiles. Driver profile and customization is when a driver adjusts the driver's seat, steering wheel position, or mirrors; the touchscreen prompts the creation of a driver profile to save these adjustments [33].

These profiles also save different preferences modified by each driver, such as climate control settings, audio preferences, and navigational instruction settings, which can be stored to the cloud and accessed across several Tesla vehicles by setting up a Tesla Profile, which means that once a driver sets up their profile, they can transfer their individual preferences to any other Tesla vehicle they drive. The Simple Section setting is a capability that changes the driver's seat and controlling wheel to help with entering and leaving the vehicle. This setting can be matched with a driver profile, taking into consideration robotized changes while entering and leaving the vehicle. For instance, when a driver moves toward the vehicle and opens it utilizing their key or the Tesla portable application, the Simple Section setting will naturally change the seat and controlling wheel to simplify it for the driver to get in and out of the vehicle. Tesla cars give a Valet mode that debilitates key capabilities and settings while the vehicle is being driven by somebody other than the proprietor [34].

The mode might be locked in by means of the touchscreen or through the portable application, and it needs a 4-digit PIN for enactment and deactivation. At the point when Valet mode is empowered, it restricts the vehicle's speed, speed increase, and admittance to capabilities like voice orders and Autopilot. This mode is valuable when the proprietor needs to loan their vehicle to somebody yet needs to limit a few elements and assurance the vehicle is driven capably. Driver profiles can be associated with keys, which infers that each key can be related with a specific driver profile. At the point when a driver opens the vehicle with their key, the vehicle will consequently pick the important profile and apply the exceptional settings related with that profile. This capability ensures that every driver's inclinations are applied relying upon the key used to open the vehicle, making it more straightforward for various drivers to have a similar vehicle without having to physically change the settings each time [35].

9.18 CONCLUSION

The development of innovative and enhanced assistive robots is changing the lives of persons with impairments and helps them with their mobility constraints. With the help of intelligent wheelchairs like the Vulcan, which supports enhanced autonomy in navigating, to multifunctional assistive robots like Toyota's Human Support Robot that can perform a wide range of duties, these technologies are boosting users' skills and freedom. Exoskeletons such as ReWalk Personal 6.0 allow persons with lower body paralysis to begin walking and standing again. Advanced prosthetic limbs like the Motion Control iLimb Ultra give more natural and precise control. Robots are additionally being developed expressly to aid the elderly, such as TUM's intelligent walker CP-Walker 2.0, which helps avoid falls. Telepresence robots like Dawn Avatar Robot give users a virtual presence to reduce isolation and engage with others remotely. And smart technologies like XanderGlasses leverage AI to assist people with visual impairments navigate their surroundings. In addition, autonomous vehicles equipped with customized assistive features can greatly expand accessible transportation options for the disabled. As research in human–robot interaction, AI, and robotics continues to accelerate, so too will assistive technologies that can enrich and empower lives. While the current landscape of assistive robots already enables life-changing applications, the future promises even more intuitive and helpful machine assistants that can fluidly work with humans as partners. The great potential of further integrating emerging technologies with a deep focus on user-centric design presages transformative advances in assistive robotics ahead.

REFERENCES

[1] V. Ahumada-Newhart *et al.*, "Evaluation of the Toyota Human Support Robot (HSR) for Social Interaction and Learning," *Int J Technol Knowl Soc*, vol. 19, pp. 21–52, Mar. 2023, doi: 10.18848/1832-3669/cgp/v19i01/21-52

[2] WHO, "Assistive Technology." World Health Organization: WHO, Mar. 2023. [Online]. Available: www.who.int/news-room/fact-sheets/detail/assistive-technology

[3] "XanderGlasses." [Online]. Available: www.xander.tech/xanderglasses

[4] W.-I. W. D. Team, "Naqi Logix Neural Earbuds: Revolutionizing Accessibility and Winning Big at CES 2024." Mar. 2023. [Online]. Available: https://www.naqilogix.com/post/naqi-logix-neural-earbuds-revolutionizing-accessibility-and-winning-big-at-ces-2024

[5] C. Bayón *et al.*, "CPWalker: Robotic Platform for Gait Rehabilitation in Patients with Cerebral Palsy," Mar. 2016, doi: 10.1109/icra.2016.7487561

[6] C. Bayón *et al.*, "Development and Evaluation of a Novel Robotic Platform for Gait Rehabilitation in Patients with Cerebral Palsy: CPWalker," *Rob Auton Syst*, vol. 91, pp. 101–114, Mar. 2017, doi: 10.1016/j.robot.2016.12.015

[7] J. Madrid, "CP-Walker 2.0: Exoskeleton for PC Patient's Mobility." Mar. 2023. [Online]. Available: https://stemcell-host.com/cp-walker-2-0-exoskeleton-for-pc-patients-mobility/

[8] R. B. van Dijsseldonk, I. J. W. van Nes, A. C. H. Geurts, and N. L. W. Keijsers, "Exoskeleton Home and Community Use in People with Complete Spinal Cord Injury," *Sci Rep,* vol. 10, p. 15600, Mar. 2020, doi: 10.1038/s41598-020-72397-6

[9] A. Zeiaee, R. Soltani-Zarrin, R. Langari, and R. Tafreshi, "Design and Kinematic Analysis of a Novel Upper Limb Exoskeleton for Rehabilitation of Stroke Patients." In *2017 International Conference on Rehabilitation Robotics (ICORR)*, pp. 759–764, Mar. 2017. doi: 10.1109/ICORR.2017.8009339

[10] C. A. Cifuentes and M. Munera, *Interfacing Humans and Robots for Gait Assistance and Rehabilitation*. Springer, 2022.

[11] A. Pino, N. Tovar, P. Barria, K. Baleta, M. Múnera, and C. A. Cifuentes, "Brain–Computer Interface for Controlling Lower-Limb Exoskeletons". In Carlos A. Cifuentes and Marcela Múnera (Eds.), *Interfacing Humans and Robots for Gait Assistance and* Rehabilitation, pp. 237–258. *Springer eBooks,* Mar. 2021, doi: 10.1007/978-3-030-79630-3_9

[12] I. Tijjani, S. Kumar, and M. Boukheddimi, "A Survey on Design and Control of Lower Extremity Exoskeletons for Bipedal Walking," *Applied Sciences*, vol. 12, p. 2395, Mar. 2022, doi: 10.3390/app12052395

[13] M. Geravand, C. Werner, K. Hauer, and A. Peer, "An Integrated Decision Making Approach for Adaptive Shared Control of Mobility Assistance Robots," *International Journal of Social Robotics*, vol. 8, pp. 631–648, Mar. 2016, doi: 10.1007/s12369-016-0353-z

[14] M. Tran, L. Gabert, S. Hood, and T. Lenzi, "A lightweight robotic leg prosthesis replicating the biomechanics of the knee, ankle, and toe joint," *Science Robotics*, vol. 7, Mar. 2022, doi: 10.1126/scirobotics.abo3996

[15] D. Stevenson, "Utah Bionic Leg in Science Robotics." Mar. 2022. [Online]. Available: www.mech.utah.edu/utah-bionic-leg-in-science-robotics/

[16] O. van der Niet, R. M. Bongers, and C. K. van der Sluis, "Functionality of i-LIMB and i-LIMB Pulse Hands: Case Report," *Journal of Rehabilitation Research & Development*, vol. 50, pp. 1123–1128, 2013, doi: 10.1682/jrrd.2012.08.0140

[17] T. Weaver, "Onshape Startup Program Spotlight: Tatum Robotics." [Online]. Available: www.onshape.com/en/blog/tatum-robotics-startup

[18] "Tatum Robotics." [Online]. Available: https://tatumrobotics.com/home/about/

[19] L. Alboul, M. Dimitrova, A. Lekova, V. G. Kaburlasos, and P. Mitrouchev, "Editorial: Emerging Technologies for Assistive Robotics: Current Challenges and Perspectives," *Frontiers in Robotics and AI*, vol. 10, Mar. 2023, doi: 10.3389/frobt.2023.1288360.

[20] S. Gallo *et al.*, "Augmented White Cane with Multimodal Haptic Feedback," *IEEE International Conference on Biomedical Robotics and Biomechatronics,* Mar. 2010, doi: 10.1109/biorob.2010.5628066

[21] "Optimus 2 RS – Electric Mobility." [Online]. Available: www.electricmobility.co.uk/products/optimus-2-rs/268/

[22] "Meyra Optimus 2 RS." [Online]. Available: www.clearwellmobility.co.uk/meyra-optimus-2-rs.html

[23] N. Aziz, N. S. Khusaini, Z. Mohamed, A. Hamid, Y. Yusof, and M. R. Aziz, "Smart Wheelchairs: A Review on Control Methods." In *2022 IEEE International Conference on Artificial Intelligence in Engineering and Technology (IICAIET)*, pp. 1–6, Mar. 2022. doi: 10.1109/IICAIET55139.2022.9936836

[24] R. Krishnamurthi, A. Kumar, S. S. Gill, and F. Xhafa, *Autonomous and Connected Heavy Vehicle Technology*. Elsevier Science & Technology, 2022.

[25] L. Shaoshan, L. Liyun, T. Jie, W. Shuang, and G. Jean-Luc, *Creating Autonomous Vehicle Systems, Second Edition*. Springer Nature, 2022.

[26] F. Group, "What Are the 6 Levels of Autonomous Vehicles?" Mar. 2023. [Online]. Available: www.faistgroup.com/news/autonomous-vehicles-levels/

[27] C. Huang, G. S. Wasson, M. Alwan, P. Sheth, and A. Ledoux, "Shared Navigational Control and User Intent Detection in an Intelligent Walker.," *National Conference on Artificial Intelligence*, pp. 59–66, Mar. 2005.

[28] C. Johnson and M. McMahon, "The Intelligent Wheelchair Project." [Online]. Available: https://web.eecs.umich.edu/~kuipers/research/wheelchair/

[29] R. C. Simpson, "How many people would benefit from a smart wheelchair?," *The Journal of Rehabilitation Research and Development*, vol. 45, pp. 53–72, Mar. 2008, doi: 10.1682/jrrd.2007.01.0015

[30] N. S. Patankar, D. A. Shaharam, D. T. Rajendra, G. S. Bhausaheb, P. S. Satish, and Y. S. Deshmukh, "IoT Based Hand Gesture Control Wheelchair for Disabled Person." In Proceedings of the Fourth International Conference on Smart Electronics and Communication (ICOSEC-2023), pp. 346–352. Mar. 2023, doi: 10.1109/icosec58147.2023.10276006

[31] S. D. Sierra M, M. Múnera, T. Provot, M. Bourgain, and C. A. Cifuentes, "Evaluation of Physical Interaction during Walker-Assisted Gait with the AGoRA Walker: Strategies Based on Virtual Mechanical Stiffness," *Sensors (Basel)*, vol. 21, p. 3242, Mar. 2021, doi: 10.3390/s21093242

[32] S. D. S. Sierra M., M. Garzón, M. Múnera, and C. A. Cifuentes, "Human–Robot–Environment Interaction Interface for Smart Walker Assisted Gait: AGoRA Walker," *Sensors*, vol. 19, p. 2897, Mar. 2019, doi: 10.3390/s19132897

[33] T. Z. Noonan, P. Gershon, B. Mehler, and B. Reimer, "Characterizing the Use of Tesla's Auto Lane Change Feature in Driver-Initiated Maneuvers," *Proceedings of the Human Factors and Ergonomics Society Annual Meeting*, vol. 66, pp. 1442–1446, Mar. 2022, doi: 10.1177/1071181322661262

[34] I. Ali, "Tesla Update 2022.24 Introduces Cloud-based Driver Profiles and Blind Spot Camera Feed Enhancements." Mar. 2022. [Online]. Available: www.teslaoracle.com/2022/07/29/tesla-update-2022-24-introduces-cloud-based-driver-profiles-and-blind-spot-camera-feed-enhancements/

[35] J. Klender, "Tesla Launches 'Profiles' for Synchronized Settings across Multiple Vehicles." Mar. 2022. [Online]. Available: www.teslarati.com/tesla-profiles-cloud-based-settings-software-update-2022-24-5/

Chapter 10

Automated solar-powered bot for identifying damage in vegetable fields using AI

Kitty Tripathi and Sushant Bhatt

10.1 INTRODUCTION

Since agriculture and related sectors continue to be one of the main sources of income for the majority of the population in rural India, India is recognized as an agricultural nation. It contributes over 17% of India's total GDP and is a crucial industry for the country's economy. India ranks second globally in the production of rice, wheat, sugarcane, groundnuts, vegetables, fruit, and jute. It is the world's greatest producer of milk, pulses, and jute. In addition, it is among the top producers of fish, poultry, cattle, and plantation crops [1].

The climate in India varies greatly for agriculture, ranging from extremely humid conditions to tropical regions that are quite dry in the southern part of the nation. On the other hand, the northern area has a moderate climate and high altitude, which leads to a large range of habitats. India plays a vital role in supplying food for a portion of the world's hunger even though the agricultural sector is producing an excess of food. As a result, India has a duty to provide food for around 190 million undernourished people [2].

Since India's agriculture is a source of innovative irrigation techniques, it now poses severe sustainability concerns, making it imperative that management of agricultural practices be improved on many fronts. The Indian agri-sector is rapidly changing and accepting automation as part of the Digital India movement, and artificial intelligence is the main technology that has been pushing the industry away from traditional methodologies and traits in recent years [3].

Considering AI as a data-driven technology, there is a lot of constantly changing data that results in ambiguity, inaccuracy, and inconsistencies. With the aid of iterative methodologies and the neural community designs associated to each difference, systems for approaching features are currently common [4, 5]. These systems are jointly investigated under the titles "deep learning" and "machine learning."

The use of robotics in agriculture might be a creative move to improve labor productivity. Robots can assist in reducing important human limits,

 DOI: 10.1201/9781003537083-10

such as the capacity to work in challenging agricultural situations, by either duplicating human talents or advancing them [6, 7].

The morphological elements of a vegetable's appearance, which are what determine its quality, are important. The most common characteristic to determine if a vegetable has been affected by weeds or not is its color. The most significant characteristic of the many elements, such as size, texture, form, and color, is color, which has a significant influence on the quality of the vegetable.

10.2 RECENT ADVANCEMENT OF THE AGRICULTURE SECTOR

AI is becoming more and more incorporated into agricultural practices to enhance different procedures. By 2022, the following developments have been made:

(i) **Precision agriculture:** in order to make exact judgments regarding planting, irrigation, fertilizing, and pest management, AI and machine learning algorithms have continued to be employed to evaluate data from many sources, including satellites, drones, and sensors. This can boost yields and optimize resource use.

(ii) **Crop disease detection:** using photographs taken by drones or cell phones, sophisticated AI-powered image recognition systems may spot symptoms of illnesses, pests, or nutritional deficits in crops. Early identification decreases crop losses and enables prompt response.

(iii) **Autonomous farming:** more autonomous farming equipment is now available thanks to advancements in robotics and AI. The ability of tractors, harvesters, and other machinery to function with little assistance from humans increases productivity and lowers labor costs.

(iv) **Weather prediction and analysis:** AI is used to anticipate extreme weather occurrences and understand complicated weather patterns. This knowledge aids farmers in preparing for bad weather and modifying their methods accordingly.

(v) **Supply chain organization:** the supply chain may be improved by AI algorithms, allowing farmers to interact with markets and customers more effectively, eliminating waste, and guaranteeing that fresh fruit reaches consumers on schedule.

(vi) **Plant breeding and geonomics:** to speed up the creation of new crop varieties with desirable features, such as drought resilience, disease resistance, and greater yields, AI can examine enormous volumes of genetic data.

(vii) **Agriculture robot swarms:** AI has made it possible to create small, collaborative robots that work in swarms rather than utilizing big, individual robots. These swarms are more effective in planting, weeding, and harvesting duties.
(viii) **Data-driven decision making:** by combining data from numerous sources, such as field sensors, satellite imaging, and historical records, AI-powered analytics' solutions assist farmers in making data-driven choices.
(ix) **Smart irrigation system:** in order to optimize irrigation schedule, reduce water wastage, and improve crop health, AI-controlled irrigation systems monitor soil moisture levels and weather forecasts.
(x) **Vertical farming and controlled environment:** in order to maximize crop development in a regulated setting, AI is employed in indoor vertical farms to regulate environmental elements, including light, temperature, humidity, and nutrients.

By utilizing artificial intelligence and technology to enhance different areas of farming and agricultural operations, smart AI-based agriculture seeks to accomplish the following main objectives:

(a) **Crop yield improvement:** increasing crop yields is one of the key objectives. Artificial intelligence (AI) algorithms examine data from many sources, including as satellite photography, drones, and sensors, to offer insights about crop health, growth trends, and the best times to sow and harvest. Farmers can make decisions based on this knowledge, which may result in improved yields.
(b) **Efficiency in resource use:** AI-driven precision agriculture helps to maximize the use of resources like water, fertilizer, and pesticides. Artificial intelligence (AI) technologies enable farmers to apply resources just where they are required, decreasing waste and environmental effect by delivering real-time data on soil moisture, nutrient levels, and insect infestations.
(c) **Cost reduction:** AI-based agriculture can help farmers cut expenses by enhancing resource efficiency and streamlining operations. Farmers may increase profitability by limiting inputs and maximizing outputs.
(d) **Environmental sustainability:** By reducing the use of pesticides, saving water, and encouraging sustainable agricultural methods, smart agriculture strives to lessen the environmental impact of agriculture. AI supports ecosystem monitoring and management to stop soil erosion, water pollution, and other detrimental effects.
(e) **Data-driven decision making:** AI systems examine a lot of data to provide farmers useful information. Making educated decisions

regarding planting, irrigation, pest management, and other crucial farming decisions is made possible as a result.

(f) **Early disease and pest detection:** image recognition and sensor technologies driven by AI can spot early symptoms of illnesses, pests, and nutritional deficits in crops. This enables farmers to take quick action to stop the spread of these problems and reduce crop losses.

(g) **Labor efficiency:** automation and robots powered by AI can take over labor-intensive, repetitive jobs, hence requiring less physical labor. In order to overcome the manpower deficit in agriculture, this is very helpful.

(h) **Plant breeding innovation:** AI speeds up the analysis of genetic data, resulting in the creation of new crop types that are more tenacious, disease-resistant, and more adapted to certain settings.

(i) **Education and information sharing:** even in remote locations, AI-powered platforms can give farmers access to expert information, best practices, and training materials, enabling them to hone their abilities and keep up with the most recent developments.

In general, the main objectives of smart AI-based agriculture are on boosting productivity, sustainability, and profitability while reducing detrimental environmental effects and tackling the difficulties that modern agriculture faces.

10.3 DIGITAL AGRICULTURE

Even though a sizable portion of the people in India still works in agriculture, there are several difficulties. Youth engagement in the industry and the labor deficit have both been worse recently [8]. The automation of Indian agriculture is also limited by a lack of resources and knowledge. Therefore, in order to boost output in terms of both quality and quantity, there is a greater demand for superior equipment and improved methods.

Agriculture methods are heavily influenced by variables such as climatic conditions, water levels, temperatures, soil fertility, weather forecasts, availability of fertilizers, usage of pesticides, presence of weeds, and harvesting techniques.

In many instances, Indian farmers use traditional methods to forecast crop yield, which depend on knowledge gained from prior experiences. However, this approach by itself may not be an effective way to forecast, as the climatic conditions frequently change dramatically as a result of the general shift in weather forecasts at the global level. With the development of technology, a new scientific method known as agro-based big data analytics may be used to address this problem [9]. Big data analysis offers the chance to examine the essential elements that are necessary for regulating agricultural productivity.

The investigation of the effects of social, economic, and political factors on the rates of success of various agricultural methods may also be done using this methodology.

One fundamental way to improve agricultural output is to expand the total amount of land that can be cultivated and is suitable for growing a specific crop. It also necessitates the adoption of numerous ethical agriculture methods, which will reduce crop damage and net operating costs.

Controlling the important elements used in Indian agriculture, such as the type and amount of fertilizer, the amount of water used and the source from which it is obtained, the quality of the seeds used in crop production, and the reduction of biotic stress brought on by weeds and pests, is necessary to increase crop yield.

There is also the practice of manual field inspection by farmers, and the removal of weed and contamination found, if any, is done manually. This method, however, is less effective and does not result in any significant limitations in support of higher crop yields, so it can be replaced by a sensor network installed in the field that can effectively understand the needs of the crop in real time [10–12].

Even though agriculture is heavily mechanized in some areas of the nation, the majority of operations in the agricultural sector are still carried out manually using traditional instruments and implementation methodologies. Machines are used either very little or not at all throughout the procedures of planting, irrigating, thinning, and pruning, weeding, harvesting, threshing, and moving the crops [13, 14]. As a result, it becomes labor-intensive and expensive, and the output from the labor force is not even excellent.

In order to equip themselves with the most recent technological breakthroughs in the industry, farmers will be able to learn about the numerous growing techniques that are employed across the world for a certain type of crop or vegetable thanks to the digital farming system [14]. In this way, farmers may employ their inherited agricultural techniques in conjunction with practical knowledge and equipment to boost output [15].

They may also become aware of their previous data, which can aid in comprehending the various circumstances and challenges and provide them with the crucial knowledge they need to make the best choice.

The historical data of any farm may be used in a variety of ways to forecast situations like the presence of weeds and pests and can hence aid in balancing the requirements of creating a profitable company by taking the following aspects into consideration.

Soil management and crop nutrition come first, followed by crop rotation, energy use, waste and pollution control systems, organization management, the agriculture site, monitoring and auditing, and crop protection. Management of wildlife and landscapes.

10.4 SOLAR ROBOT

A simple, economical, and very effective technique for using light to drive soft robots is presented by the authors in [16]. The method allows for the long-range excitation of soft robots using collimated light or sunshine, making it applicable in locations without access to electric or pneumatic sources. We show lifting and grabbing everyday objects using two separate soft robot prototypes and a low-level control system. These characteristics provide fresh possibilities for soft robot and even stiff robot applications. Additionally, our suggested method may be applied at scales (from micro to macro) where it is hard to apply tethered soft robots.

Although solar-powered electric farm tractors and agricultural robots are promoted as environmentally friendly and sustainable alternatives to conventional machines that use fossil fuels to perform agricultural tasks, there are still some difficulties with these contemporary sustainable farm vehicles that are seen as the main obstacles to their widespread use [17].

The author of [18] mentioned how the development of agricultural robots may be highly helpful in resolving labor concerns and boosting farm output by 30% while remaining affordable. Software called Solid Works and Ansys were used to create and analyze the robot, respectively.

The surveillance robot created and tested in the study has been effectively used in an agricultural-like context with all-terrain contours and the ability to avoid obstacles, according to the author in [19] and the quickest charging time occurred in bright, high-irradiance conditions and was 2 hours.

The author's goal for this research, as stated in [20], is to create a solar powered robot for various agricultural uses such as spraying fertilizer on fields. Furthermore, trimming the extra grass around the crops, as well as the right manner to spread seeds and plant the plants in large or small areas. This robot is made out of a solar panel to charge the battery, four motors, various sensors, a pixy2 camera to analyze photos and help the robot travel along the tracking line, and the most significant element of the robot is an Arduino UNO. Finally, the robot was tested in agricultural land and found to be capable of achieving the intended objectives.

The author said in [21] that the FLC design is also designed to forecast the charging source of the robot based on the light sensor linked to the charging system. On a bright day, the robot may get electricity straight from the sun, but on a foggy day, it can draw power from the battery. MobotSim is used to model the robot's mobility as it goes from one location to another while gathering the agricultural crop. The simulations undertaken in this work suggest that a solar-powered mobile manipulator is feasible for use as a harvesting robot in agriculture.

10.5 MECHANICAL DESIGN OF AN AGRICULTURAL ROBOT

The deployment of various individual physical sensors in the field is time-consuming, costly, and difficult to maintain, so an unmanned portable robot (Figure 10.1) equipped with the necessary sensors can be used to effectively monitor the growth time course of vegetables [22, 23, 24].

This sort of solar power unit is utilized in the agricultural area for pesticide spraying and irrigation. When not in use in the agricultural field, these units may be brought back home and used to power household appliances by electric vehicle or tractor.

A solar tracker directs the solar panels toward the sun. Solar trackers are used to compute the analogy values generated by LDR (light-dependent resistors). Our prototype includes a dual-axis solar tracking system with automated and human control. The automated control system uses a single-axis rotation system, which rotates the panel

Figure 10.1 Carrier trolley setup of the robot.

horizontally (left to right), whereas the dual-axis rotation system turns the panel up and down.

This system connects five solar panels in series to create 3 V and 0.25 Amp electricity. With the aid of a 12 V DC source, this voltage and current are boosted to 13.5 V and 0.1 Amp. Both AC and DC power supplies are employed in our model, which may be switched in a specific option from the control panel, and the back view of the module is displayed in the image below.

This system employs the ESP32-CAM, which has three GND and two power pins. The ESP32-CAM is powered by 5 V and provides a 3.3 V output. When connected to a 3.3V source, the same output reading of 5 V is achieved. The ESP module connects to a pre-configured Wi-Fi network. By accessing the URL in a browser, we may observe the outputs of live visuals immediately from the camera.

To fulfill the goal of distinguishing between harmful and healthy vegetable conditions in the field, the robot's design includes the following aspects (Figure 10.2):

1. The robot's exterior frame is a wooden construction weighing approximately 3.5 kg.
2. The panels are fixed on the top at a height of 80 mm using a fold and unfold mechanism comprised of an ejecting plate and a plastic frame supported on the wooden structure by a wooden slab.
3. The unfolding of solar plates is controlled by a set of pushbuttons that are manually operated and positioned on either side of vertical folding segments.
4. A DC geared motor with a 5 V operating voltage controls the dual-axis tracking mechanism of the solar plate modules. As illustrated in Figure 10.2, this motor is supported by a slender rod with springs connected to it for height adjustment. The graphic depicts the horizontal and vertical motion of the panel assembly, which is controlled by a geared DC motor. This system also includes light-dependent resistors installed throughout the length and width of the solar panel array, which will assist the system in determining the point of greatest light intensity and aligning accordingly. This method improves power generating efficiency by roughly 30% over a static solar panel system.
5. As seen in Figure 10.3, this building has five solar panels placed on it. The panels used are monocrystalline and measure 158 mm × 158 mm 0.25 mm. The diagonal length is 212 mm ± 0.25 mm, and the panel thickness is 190 m ± 0.30 m. The module features a positive electrode width of 0.7 mm with a 116 root fine grid and a blue silicon nitride anti-reflective coating, and a back electrode width of 1.8 mm with an anti-reflective layer.

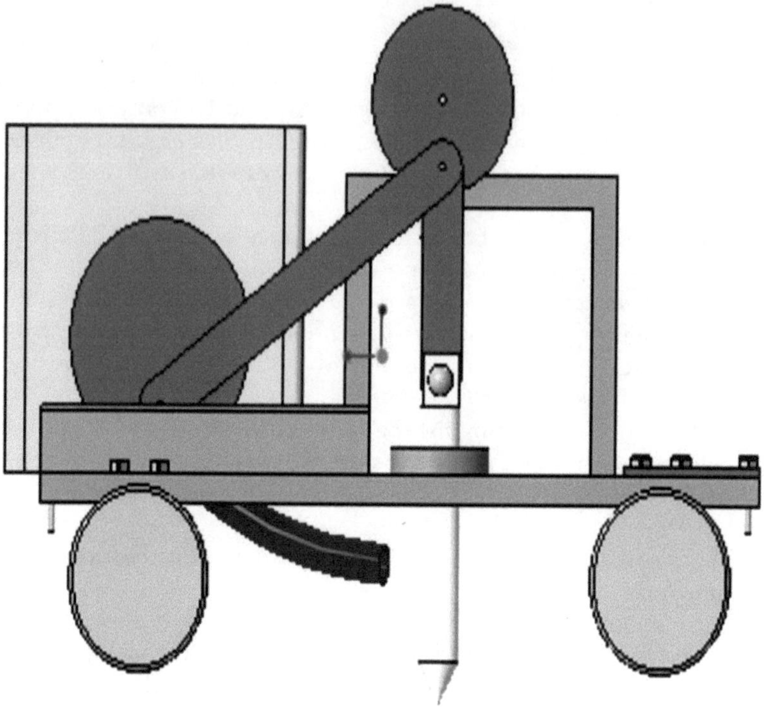

Figure 10.2 CAD design of an agriculture robot.

6. All of the solar panels are linked in series, increasing the total generating voltage to around 3.00 V and the generating current to around 250mA. Because it is insufficient for charging a battery, a DC-DC step up module (XL6009) is utilized to raise the voltage to 13.5 V DC.
7. The robot's carrier trolley is a four-wheeled device with a DC geared motor that controls the robot's mobility in all directions. The wheels utilized are constructed of plastic and are 70 mm in diameter and 20 mm in width. The shaft hole measures 6 mm in diameter.
8. This robot's main controller is the Arduino Uno, which comprises a physical programmable circuit board as well as an integrated development environment. It is a low-cost, commonly used microcontroller with enough computing capability for the job. The assembly includes four pushbuttons, four LDRs, and two toggle switches (T1 and T2) for manual control and mode switching, respectively. When the toggle T1 is set to LOW, manual mode is activated, and the four pushbuttons control the orientation of the solar panels. When T1 is

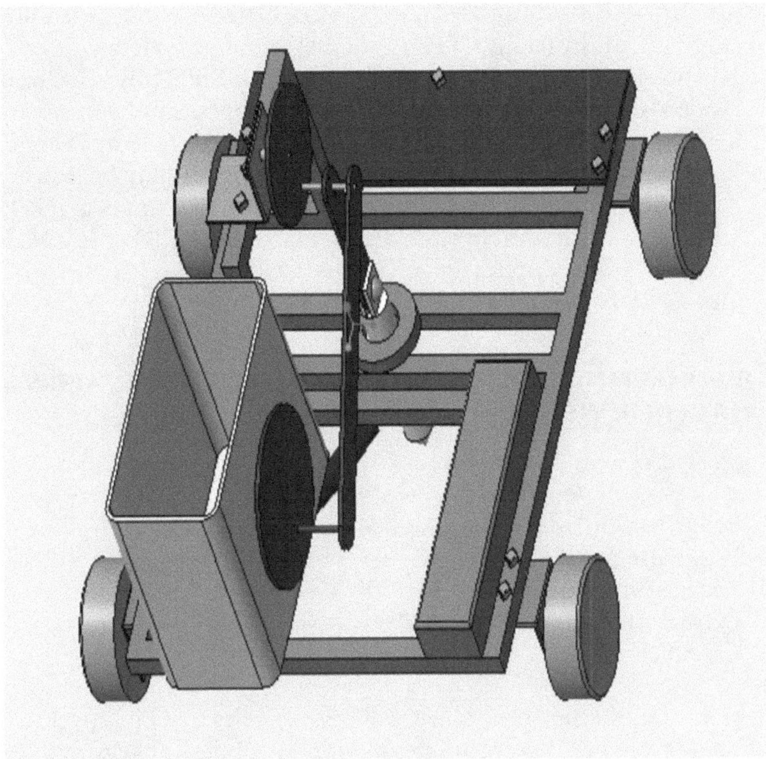

Figure 10.3 CAD design of an angled view of the robotic body.

turned on, automated mode is activated, and the program runs as follows:

9. (a) The analog readings from four LDR sensors (top-right, top-left, bottom-right, bottom-left) were read by the microcontroller.
   (b) The microcontroller then computes average values for top-bottom, left-right, i.e., avgt_op, avg_left, avg_bottom, and avg_right.
   (c) It will now calculate the difference between elevation and horizontal difference using the difference of average values.
   (d) If (horizontal difference) = 10, the left-right motor turns the panel left.
   (e) If (horizontal difference) > 0, the left-right motor rotates the panel right.
   (f) Likewise, if (difference elevation) = 10, the top-down motor rotates the panel down.
   (g) If (difference elevation) is greater than zero, the top-down motor rotates the panel top.

10. A video camera module The ESP32-CAM development board is utilized, which has an ESP32-S chip, an OV2640 camera, a micro SD card slot, and various GPIOs for connecting peripherals.
11. For uploading picture data to the cloud an ESP8266 is utilized as a NodeMCU. The NodeMCU module is an open-source firmware and development board used for Internet of Things (IoT) applications. It is powered by an Espressif System Wi-Fi SoC. It may be powered by a USB port or a 3.3 V regulated source. The CPU supports RTOS and works at a configurable clock frequency of 80 MHz to 160 MHz. To store data and applications, NodeMCU features 128 KB of RAM and 4 MB of Flash memory.

10.6 EVALUATING THE QUALITY OF VEGETABLES USING MACHINE VISION

The suggested system architecture is divided into five steps:

(i) Image for input
(ii) Image pre-processing
(iii) Image segmentation
(iv) Extraction of features
(v) Classification

In the picture gathering procedure, robotic vehicles are employed to capture photos of veggies. Relevant information for undesirable growth for the estimate of weeds is acquired from these collected photographs and then utilized to develop the data collection. These data are currently being used in picture pre-processing.

The image pre-processing methods amplify the picture data. This approach eliminates the disinclined misrepresentations and expands the picture characteristics. This is essential for the creation and processing of a relevant image in addition to the indigenous image. Local pre-processing, in addition to pixel pre-processing, can be used to complete the main characteristics and assessment of vegetables in image pre-processing [25, 26].

When the pre-processing of image is completed, the process of image segmentation is in process. In image segmentation, the digital image is converted into well-defined areas. In the evaluation of fruit, image segmentation is used to separate the background for processing the important area. In further image analysis, the original segmentation is definite. The performance of the classifiers can be reduced with the help of improper segmentation. The different types of method approaches include *k*-means clustering and fuzzy *c*-means, which can be used for segmentation [27, 28, 29].

After the features are extracted, they can then be used for other analyses. The classification of vegetables along with the image clarification are considered in the extraction feature. The morphological and textural features are used for the analysis of the growth of weed. The simulated assumptions of human prospects are helpful in the calculation of the presence of damage or weed. The image of vegetables can establish a set of information or feature like texture, color, size, and shape. These sets of information may be used in training sessions. After these training sets, the neural network algorithm is used to make a decision [30].

The classification model is used to determine the category or class of data using system training to extract the required characteristics. The classification algorithm's primary steps are as follows:

(i) The classifier has been initialized.
(ii) A fit procedure for the model generates the trained label "l" for a given training data "*t*."
(iii) The trained model is used to predict the label for any unlabeled data received.
(iv) The model's error rate is then calculated.

Sensor readings, as well as the color channels for our system's various color spaces, control the dispersion of samples in the attribute space. The examination for the color channels that become more relevant for the classification and is necessary for the analysts to find the selection of border between the two classes that are vegetable health state and damaged condition. Features that are either noisy or irrelevant might exacerbate classification issues, resulting in poor performance even when the classification is correct [31]. Manually or with the assistance of feature extraction tools, a subset of the k-most relevant characteristics can be selected. It will help in the removal of superfluous and irrelevant material.

The produced k-dimensional feature space is utilized for classification instead of the original d-dimensional feature space [32]. This is suitable for aiding in the identification of a criterion for distinguishing between the two classes. In contrast to feature extraction strategies, which attempt to create a new characteristic space, characteristic choice methods aim to pick a subset of vectors that cover a feature subspace that permits class. Unsupervised feature selection techniques rely solely on the underlying sample distribution, whereas supervised feature selection strategies use class labels to identify the most relevant characteristics [33]. The computationally costly job of picking the subset of the most important features is a combinatorial challenge. The method used to rate or choose features may have an impact on how successfully an algorithm identifies a good enough feature subset.

Several considerations can be used to identify the aim of feature selection and extraction. $X = x1, x2, \dots xN, xi\ Rd$. The feature extraction and selection operations produce a new set *X*' = *x*'1, x'2 ,..., *x*'*k*, where *x*'*i* is the new dimension of the feature vectors and *k d* seems to be a new vector element.

Evenness, variance, mean, standard deviation, RMS, inverse difference moment, kurtosis, and skewness are statistical variables used to calculate first-order probability. The histogram of pixel intensity is used in this scenario to determine statistical properties of the defective portion. Color characteristics derived from the mean and standard deviation of RGB values are also included.

Textural variables such as disparity, energy, entropy, consistency, and correlation are used to calculate second-order probability. The distribution of gray scale in the gray image approximates the image's texture.

Following the extraction of 108 characteristics, a subset of 25 features is chosen, which comprises mean, standard deviation, homogeneity, entropy, variance, axis length, RMS, inverse difference moment, coarseness, homogeneity, correlation, centroid, and orientation.

10.7 CONCLUSION AND FUTURE SCOPE

This paper proposes a solar-powered robotic cart that may assist in taking images of various phases of vegetables growing in the field via a feed from a camera module installed on the robotic vehicle. This vehicle is low-cost and unmanned, and it will be used to monitor the status of veggies in the vegetable field [34]. The picture thus acquired is utilized to detect the damaged vegetable by categorizing the image into two classes: healthy condition and unhealthy or damaged state of the vegetable at the start of the weed's development. The most essential element retrieved by the picture to identify and classify the stage of the vegetable is its color.

The artificial neural network approach was used to train it. Manual segmentation and sorting were performed prior to characteristic segmentation and identification. Until recently, morphological characteristics have been extracted using various machine learning approaches such as SVM, ANN, *K*-means clustering, KNN, and fuzzy logic. Some researchers are currently applying deep learning algorithms and a fusion of color and textural elements of crop, vegetable, or fruit images to achieve a better degree of accuracy [35, 36]. Precision, accuracy, sensitivity, specificity, FPR, and FNR are used to calculate classification performance. Using SVM, the greatest accuracy achieved for *k*-fold = 5/10/15 is 90.12%, 93.64%, and 94.48%, respectively.

As a result, deep learning algorithms with several characteristics merged for classification may be used in the future to increase accuracy [37]. Fusion statistics, such as classifier combination procedures, on the other hand, may enhance the accuracy of creating decision demarcation between two classes. The solar power robotic vehicle's future potential is to classify live image feeds and, if a damaged vegetable or weed growth is detected, it can make the decision to clean the weed and sprinkling the necessary fertilizers via a fertilizer bottle mounted on the unmanned vehicle.

Future study might focus on connecting different soil nutrient sensors with Arduino UNO and evaluating the findings to achieve proper and better results, which would further minimize production costs, water and fertilizer use. Data collecting from diverse farmlands and data analysis using data mining techniques would be appropriate for agriculture. Big data analysis may also be employed to get the desired result. The results' data may be given to farmers and agro-companies to determine their seed, fertilizer, and water needs.

However, there are obstacles to overcome, such as guaranteeing the accuracy and dependability of AI algorithms, dealing with various field settings, and resolving any possible privacy concerns associated to data collecting. Overall, the notion has enormous potential to transform current agriculture and contribute to greater food production and sustainability.

As the importance of sustainable and technology-driven agriculture grows, there is a rising demand for new solutions such as AI-powered automated bots. This can attract investment and research resources to help the technology evolve and improve.

REFERENCES

[1] Food and Agriculture Organization, "India at a glance", January 2022(Online), Available: www.fao.org/india/fao-in-india/india-at-a-glance/en/ [Accessed: January 3rd 2022]

[2] World Internet Usage and Population Statistics, 3rd July 2021(Online), Available: www.internetworldstats.com/stats.htm [Accessed: December 12th 2021]

[3] B. A. King, T. Hammond, and J. Harrington, "Disruptive technology: Economic consequences of artificial intelligence and the robotics revolution", *Journal of Strategic Innovation and Sustainability*, Volume 12, no. 2, 2017. https://doi.org/10.33423/jsis.v12i2.801

[4] J. M. McKinnon, and H. E. Lemmon, "Expert systems for agriculture, Computers and Electronics in Agriculture", Volume 1, no. 1, 1985, pp. 31–40, ISSN 0168-1699. https://doi.org/10.1016/0168-1699(85)90004-3

[5] J. L. Ruiz-Real, J. Uribe-Toril, J. A Torres Arriaza, and J. Valenciano de Pablo, "A Look at the past, present and future research trends of artificial intelligence in agriculture". *Journal of Agronomy*, Volume 10, no. 11, p. 1839, 2020.

[6] K. Liakos, P. Busato, D. Moshou, S. Pearson, and D. Bochtis, "Machine learning in agriculture: A review". *Sensors*, Volume 18, no. 8, p. 2674, 2018. http://doi.org/10.3390/s18082674

[7] F. A. Auat Cheein, and R. Carelli, "Agricultural robotics: Unmanned robotic service units in agricultural tasks", *IEEE Industrial Electronics Magazine*, Volume 7, no. 3, pp.48–58, 2018. doi: http://doi.org/10.1109/MIE.2013.2252957

[8] T. Mahindru, "Role of Digital and AI Technologies in Indian Agriculture: Potential and way forward", *Niti Aayog, Government of India*, September 2019.
[9] J. W. Jones, J. M Antle, B. Basso, K. J. Boote, R. T. Conant, I. Foster, H. Charles, J. Godfry, M. Herrero, R. E. Howitt, S. Janssen, B. A. Keating, R. Munoz-Carpena, C. H. Porter, C. R. Zweig, and T. R. Wheeler, "Toward a new generation of agricultural system data, models, and knowledge products: State of agricultural systems science," *Journal of Agricultural Systems*, Volume 155, pp. 269–288, 2017.
[10] United Nation Global Compact, "Digital Agriculture: Feeding the future", Available: http://breakthrough.unglobalcompact.org/disruptive-technologies/digital-agriculture/. [Accessed: November 12th 2021]
[11] Y. Edan, S. Han, and N. Kondo, "Automation in Agriculture" In: Yael Edan, Shufeng Han and Naoshi Kondo D. (Eds.), *Springer Handbook of Automation*, pp. 1095–1128. Springer, Berlin, Heidelberg, Germany, 2009. https://doi.org/10.1007/978-3-540-78831-7_63
[12] R. R. Shamshiri, F. Kalantari, K. C. Ting, K. R. Thorp, I. A. Hameed, C. Weltzien, D. Ahmad, and Z. Shad, "Advances in greenhouse automation and controlled environment agriculture: A transition to plant factories and urban agriculture," *International Journal of Agricultural & Biological Engineering*, Volume 11, no. 1, pp. 1–22, 2018. https://doi.org/10.25165/j.ijabe.20181101.3210
[13] S. M. Pedersen, S. Fountas, H. Have, and B. S. Blackmore, "Agricultural robots – system analysis and economic feasibility," *Precision Agriculture*, Volume 7, pp. 295–308, 2006.
[14] V. Saiz-Rubio, and F. Rovira-Más, "From smart farming towards agriculture 5.0: a review on crop data management," *Agronomy*, Volume 10, p. 207, 2020.
[15] T. Moysiadis, G. Adamides, A. Stylianou, N. Zotos, M. Giannakopoulou, and G. Alexiou, "Use of IoT technologies for irrigation and plant protection: The case for Cypriot fruits and vegetables". In: D. Bochtis, C. Achillas, G. Banias, and M. Lampridi (Eds.), *Bio-Economy and Agri-Production: Concepts and Evidence*, pp. 123–145. Academic Press, London, 2020.
[16] S. M. Mirvakili, A. Leroy, D. Sim, and E. N. Wang, "Solar-driven soft robots." *Advanced Science*, Volume 8, p. 2004235, 2021. https://doi.org/10.1002/advs.202004235
[17] S. Gorjian, H. Ebadi, M. Trommsdorff, H. Sharon, M. Demant, and S. Schindele, "The advent of modern solar-powered electric agricultural machinery: A solution for sustainable farm operations", *Journal of Cleaner Production*, Volume 292, p. 126030, 2021, ISSN 0959-6526, https://doi.org/10.1016/j.jclepro.2021.126030
[18] M. Poojari, H. Hanumanthappa, C. D. Prasad et al., "Computational modelling for the manufacturing of solar-powered multifunctional agricultural robot." : *International Journal on Interactive Design and Manufacturing*, 2023. https://doi.org/10.1007/s12008-023-01291-y
[19] T. Dewi, R. Sukwadi, and M. B. Wahju, "Design and performance of solar-powered surveillance robot for agriculture application." *Kinetik: Game Technology, Information System, Computer Network,*

Computing, Electronics, and Control, Volume 8, no. 3, 2023. https://doi.org/10.22219/kinetik.v8i3.1722

[20] S. M. Kesavan, K. S. Al Mamari, and N. S. M. Raja, "Solar powered robot for agricultural applications," *2021 International Conference on System, Computation, Automation and Networking (ICSCAN), Puducherry*, India, 2021, pp. 1–5, doi: 10.1109/ICSCAN53069.2021.9526436.

[21] F. Septiarini, T. Dewi, and N.A. Rusdianasari, "Design of a solar-powered mobile manipulator using fuzzy logic controller of agriculture application", *International Journal of Computer Vision and Robotics*, Volume 12, no. 5, pp. 506–531, 2022. https://doi.org/10.1504/ijcvr.2022.125356

[22] Y. Edan, S. Han, and N. Kondo, "Automation in agriculture." In: Shimon Y. Nof, (Ed.), *Springer Handbook of Automation*, pp. 1095–1128. Springer, Berlin, Heidelberg, 2009.

[23] A. Durand-Petiteville, E. Le Flecher, V. Cadenat, T. Sentenac, and S. Vougioukas, "Tree detection with low-cost three-dimensional sensors for autonomous navigation in orchards." *IEEE Robotics and Automation Letters*, Volume 3, pp. 3876–3883, 2018.

[24] W. Ji, V. I. Adamchuk, S. Chen, A. S. Mat Su, A. Ismail, Q. Gan, Z. Shi, and A. Biswas, "Simultaneous measurement of multiple soil properties through proximal sensor data fusion: a case study." *Geoderma*. Volume 341, pp. 111–128, 2019.

[25] K. Berger, J. Verrelst, J. B. Féret, Z. Wang, M. Wocher, M. Strathmann, M. Danner, W. Mauser, and T. Hank, "Crop nitrogen monitoring: recent progress and principal developments in the context of imaging spectroscopy missions." *Remote Sensing of Environment*, Volume 242, p. 111758, 2020.

[26] L. García, L. Parra, J. M. Jimenez, J. Lloret, and P. Lorenz, "IoT-based smart irrigation systems: An overview on the recent trends on sensors and iot systems for irrigation in precision agriculture." *Sensors (Basel*), Volume 20, p. 1042, 2020.

[27] K. Seng, L.-M. Ang, L. Schmidtke, and S. Rogiers, "Computer vision and machine learning for viticulture technology". *IEEE Access*, pp. 67494–67510, 2019. https://doi.org/10.1109/ACCESS.2018.2875862

[28] P. Skolik, C. L. M. Morais, F. L. Martin, and M. R. McAnish, "Determination of developmental and ripening stages of whole tomato fruit using portable infrared spectroscopy and Chemometrics", *BMC Plant Biology*, Volume 19, no. 236, 2019. https://doi.org/10.1186/s12870-019-1852-5

[29] D. Du, J. Wang, B. Wang, L. Zhu, and X. Hong, "Ripeness prediction of postharvest kiwifruit using a MOS E-Nose combined with chemo metrics", *Sensors Volume,* Volume 19, no.2, p. 419, 2019. https://doi.org/10.3390/s19020419

[30] S. Cárdenas-Pérez, J. Chanona-Pérez, J. V. Méndez-Méndez, G. Calderón-Domínguez, R. López-Santiago, M. J. Perea-Flores, and I. Arzate-Vázquez, "Evaluation of the ripening stages of apple (Golden Delicious) by means of computer vision system", *Biosystems Engineering*, Volume 159, pp. 46–58, 2017. https://doi.org/10.1016/j.biosystemseng.2017.04.009

[31] W. Castro, J. Oblitas, M. De-la-Torre, C. Cotrina, K. Bazán, and H. Avila-George, "Using machine learning techniques and different color spaces for the classification of Cape gooseberry (Physalis peruviana L.) fruits according

to ripeness level", *PeerJ Preprints* Volume 7, p. e26691v2. https://doi.org/10.7287/peerj.preprints.26691v [Accessed: 3rd December 2021].

[32] U. Abdulhamid, M. A. Abdulhamid, and S. Daniel, "Detection of soya beans ripeness using image processing techniques and artificial neural network", *Asian Journal of Physical and Chemical Sciences*, Volume 5, no. 10, 2019. https://doi.org/9734/ajopacs/2018/39653

[33] O. S. Taghadomi-Saberi, Z. Emam-Djomeh Mahmoud, and Kh. Faraji-Mahyari, "Determination of cherry color parameters during ripening by artificial neural network assisted image processing technique", *Journal of Agricultural Science and Technology*, Volume 17, pp. 589–600, 2015. https://doi.org/10.13140/2.1.2049.6006

[34] N. Chandra, B. Tudu and C. Koley, "A Machine Vision-Based Maturity Prediction System for Sorting of Harvested Mangoes", *Instrumentation and Measurement, IEEE Transactions,* Volume 63, pp. 1722–1730, 2014. https://doi.org/10.1109/TIM.2014.2299527

[35] M. S. Uddin, and J. Bansal, "Introduction to computer vision and machine learning applications in agriculture". In: Mohammad Shorif Uddin (Ed.), *Algorithm for Intelligent System*, pp. 1–8. Springer, Berlin, Heidelberg, Germany, 2021. https://doi.org/10.1007/978-981-33-6424-0_1

[36] X. C. Ting Yuan, "Information acquisition of cucumber fruit in greenhouse environment based on nearing infrared image", *Information Processing in Agriculture*, Volume 7, no. 1, pp. 1–19, 2020. https://doi.org/10.1016/j.inpa.2019.09.006

[37] Q. Zhang, S. Chen, and T. Yu, "Cherry recognition in natural environment based on the vision of picking robot", *IOP Conference Series Earth Environment Science*, Volume 61, no. 1, pp. 1–7, April 2017. https://doi.org/10.1088/1755-1315/61/1/012021

Chapter 11

Robots in space exploration

Prabhu Jyot Singh and Sai K Lakkaraju

11.1 INTRODUCTION

Nowadays technology has become increasingly vital in the field of space exploration, opening up opportunities for research and development. Robotics has emerged as a component in this domain. Robots have been part of space exploration for quite some time. Recent advances in technology have greatly increased their capabilities. The field of artificial intelligence (AI) has played a role in making robots more efficient and effective for space exploration.

Currently, many research organizations and space agencies are prioritizing the integration of robotics to ensure space missions. The utilization of arms, autonomous operations in space, and employing robots for repairing satellite centers have effectively addressed challenges faced during space exploration endeavors and have contributed to mission success. These cutting-edge technologies are proving to be highly capable and efficient in resolving a multitude of issues encountered during space missions. However, further research is still required to enhance the success rate by integrating AI with robotics.

This chapter digs into the use of robotics in space exploration, covering various applications that are currently employed in space missions. Many space agencies around the world are already exploring the capabilities of robotics to provide broader objectives in their missions, and ongoing research focuses on providing efficient ways by which space missions can be completed more efficiently with the help of robotics. In this chapter, a historical overview of the use of robotics in satellites and space missions is discussed. The key milestones and classification of various space robots are also covered in this chapter. It is required to understand the utilization of space robotic applications by which completely enhanced knowledge of space robots can be provided as such and we covered various applications of robotics in space exploration and addressed the ethical considerations surrounding their use in this chapter.

DOI: 10.1201/9781003537083-11

The remainder of the chapter is organized as follows: Section 11.2 provides a historical overview, Section 11.3 covers milestones and key developments of space robots, and Section 11.4 elaborates on planetary exploration. Asteroid and comet exploration are discussed in Section 11.5, while Sections 11.6 and 11.7 cover satellite servicing and repair, as well as space mining and resource utilization, respectively. The collaboration of human–robot teams is explained in Section 11.8. Ethical and environmental considerations are addressed in Section 11.9, and Section 11.10 focuses on future prospects and emerging technologies. Finally, the chapter is concluded in Section 11.11.

11.2 HISTORICAL OVERVIEW

The era of robotic space exploration began in the middle of 20th century with the introduction of artificial satellites. On October 4, 1957, the Soviet Union's first satellite Sputnik 1 rose into the sky. This event started the space age by ushering into the era of human-made objects circling the Earth. This event also sparked the space race rivalry between the United States and the Soviet Union. Launched on September 12, 1959, Luna 2, the successor in the Soviet Union's Luna program, ventured to the Moon. It was the first human-made object that reached the moon. Luna 2 successfully collided with the Moon, provided important data about the lunar environment. Luna 2 achieved a significant milestone in interplanetary exploration.

The 1960s saw the launch of various robotic missions, including NASA's Mariner series (1962–1971) [1] and the Soviet Union's Venera series (1961–1984) [2]. Launched in 1962, NASA's Mariner 2 successfully completed its first interplanetary mission by conducting a flyby of Venus. This exercise provided valuable insight into the planet's atmosphere and surface conditions. Launched in 1966, the Soviet spacecraft Luna 9 achieved an amazing feat of the first soft landing on the Moon, capturing, and beaming back photographs from the lunar surface.

In the 1960s, the early stages of robotic space exploration, a notable challenge emerged: the lack of reliable communication systems [3, p. 5]. Ensuring a stable communication link between the Earth and the robotic spacecraft proved to be a key element of mission success. Another challenge was navigation accuracy. Calculating the trajectory and making precise course corrections was essential to reaching the intended goal in the vastness of space. In addition, the payload capacity of early spacecrafts was limited, so making small instruments to fit was a challenge. Scientists and engineers had to create compact and lightweight scientific instruments. They are somewhat successful in addressing all these challenges in current spacecrafts.

11.2.1 Milestones and key developments in robotic space exploration

NASA's Surveyor program paved the way for manned lunar landings. Launched in 1966 Surveyor 1 [4] successfully made a soft-landing on the Moon, demonstrating the feasibility of human landing.

In the 1970s NASA's Viking program [5] consisted of two landers and two orbiters, providing the first successful soft landings on Mars. These missions have greatly expanded our understanding of the Martian surface and atmosphere.

In 1977, NASA's Voyager 1 [6] and Voyager 2 [7] space crafts conducted long-duration flybys of Jupiter, Saturn, Uranus, and Neptune, producing detailed images and data of these distant planets and their moons. Voyager 1 reached interstellar space in August 2012. Voyager 2 reached interstellar space in 2018.

Launched in 1999, NASA's Stardust [8] and Launched in 2004, ESA's Rosetta [9] collected samples from comets and asteroids. They provided insights into the composition of these comets and asteroids.

NASA's Mars rovers namely Sojourner [10], Spirit [11], Opportunity [12], Curiosity [13], and Perseverance [14], have made significant discoveries about the Martian surface, geology, and potential habitability, furthering our understanding of the Red Planet.

These milestones and developments highlight the progression of robotic space exploration, from the early challenges faced to the remarkable achievements that have expanded our knowledge of the solar system and the universe beyond.

11.3 TYPES OF SPACE ROBOTS

Space robots can be classified into several groups based on their functionality. The following subsections discuss some of the space robots namely rovers, landers, and space telescopes and observatories in detail. Furthermore, some futuristic technologies such as humanoids, and swarm robots are discussed briefly.

11.3.1 Rovers

Rovers are crafted for traversing the surfaces of planets and other celestial bodies. They are fitted with either wheels or tracks. These robots play a pivotal role in planetary exploration. Rovers come equipped with an array of scientific instruments, enabling them to analyze surfaces, conduct experiments, and capture captivating images. Rovers can be operated remotely from Earth or work independently.

The Soviet Union took the lunar lead by dispatching a series of lunar rovers called Lunokhods to the Moon. Both Lunokhod 1 and Lunokhod 2, robotic marvels of the early 1970s, tirelessly explored the lunar terrain [15]. China, too, has left its rover mark in the lunar sands through its Chang'e program. The Chang'e-3 mission [16], featuring the Yutu rover, touched down on the Moon in 2013. Undeterred, China continued its lunar odyssey with the Chang'e-4 mission, which introduced the Yutu-2 rover to explore the Moon's far side [17]. India's space agency, ISRO, made strides by deploying the Pragyan rover in the Chandrayaan-3 mission [18], bouncing back after a challenging landing with the Vikram lander during Chandrayaan-2 [19]. NASA, not to be outdone, is gearing up for lunar exploration with the Volatiles Investigating Polar Exploration Rover (VIPER) [20], set to hunt for ice and other potential resources on the Moon in late 2024.

Turning our gaze to Mars, the trailblazer in planetary roving is the Sojourner rover, a pioneer in NASA's Mars Pathfinder mission [10]. Launched on December 4, 1996, the Pathfinder successfully touched down on Mars on July 4, 1997. Subsequent Mars rovers, including Spirit, Opportunity, Curiosity, and Perseverance, have delved into Martian soils, scrutinized samples, and unraveled the planet's geological mysteries. China, mirroring its lunar success, achieved a Martian milestone with the Tianwen-1 mission, landing the Zhurong rover on Mars in May 2021 [21]. This triumph solidified China's position as the second country, after the United States, to operate a rover on the Martian surface.

11.3.2 Landers

Landers are stationary robotic platforms designed to land on a celestial body and remain in one location. Unlike rovers, landers do not move after landing. They typically carry scientific instruments to study the atmosphere, surface, and other properties of the celestial body. Landers have been used in missions to the Moon, Mars, and other planets and moons in our solar system as shown in Table 11.1.

11.3.3 Space telescopes and observatories

Space telescopes are placed in orbit around Earth or other celestial bodies. They are equipped with powerful telescopic instruments to observe distant objects in space. Space telescopes like the Hubble Space Telescope [22] capture high-resolution images of galaxies, stars, planets, and other astronomical phenomena. They provide valuable data for astronomers and scientists studying the universe.

Observatories in space can include a variety of instruments beyond telescopes, such as spectrometers and detectors. These observatories are specifically designed to capture different wavelengths of electromagnetic

Table 11.1 Types of landers

Name of the lander	*Year*	*Destination*	*Country*	*Result*
Luna 2	1959	Moon	USSR	Successful
Surveyor 1	1966	Moon	USA	Successful
Luna 9	1966	Moon	USSR	Successful
Apollo Lunar Module	1969	Moon	USA	Successful
Viking 1	1976	Mars	USA	Successful
Viking 2	1976	Mars	USA	Successful
Venera Series (Multiple Missions)	1961–1984	Venus	USSR/Russia	Successful
Mars Program (Multiple Missions)	Various	Mars	USSR/Russia	Various (Mixed Success)
Phobos Program (Multiple Missions)	1988	Mars' Moon Phobos	USSR/Russia	Contact Lost (Partial Data)
Huygens	2005	Titan (moon of Saturn)	ESA (EU)	Successful
Phoenix	2008	Mars	USA	Successful
Chang'e-3 Lander	2013	Moon	China	Successful
Schiaparelli	2016	Mars	ESA (EU)	Contact Lost (Descent Phase)
Chang'e-4 Lander	2018	Moon	China	Successful
Vikram (Chandrayaan-02)	2019	Moon	India	Contact Lost
Tianwen-1 Lander – Zhurong	2021	Mars	China	Successful
Vikram (Chandrayaan-03)	2023	Moon	India	Successful

radiation, allowing scientists to study cosmic phenomena across the electromagnetic spectrum.

The first telescope/observatory known as the orbiting Astronomical Observatory (OAO) 2 and nicknamed stargazer was launched in December 1968 [23]. Several space telescopes and observatories have been commissioned and decommissioned since then. Currently there are less than ten in operation as listed below.

Launched in 1990 the **Hubble Space Telescope (HST)** [22] provides observations across a wide range of wavelengths using 2.4-meter diameter mirror. On November 23, 2023 it went into safe mode due to an ongoing gyroscope issue.

Launched on July 23, 1999, **Chandra X-ray Observatory** [24, 25] observes X-rays from high-energy celestial objects. It uses high-resolution mirrors, X-ray telescope, advanced CCD imaging spectrometer (ACIS), aspect camera, and a high-resolution camera.

Launched on June 11, 2008, **Fermi Gamma-ray Space Telescope** [26] studies gamma rays from various cosmic sources. It uses a large area telescope (LAT), which observes gamma rays in the energy range from 8000 electron volts (8 keV) to greater than 300 billion electron volts (300 GeV).

Launched on April 18, 2018, **Tess (Transiting Exoplanet Survey Satellite)** [27] searches for exoplanets using the transit method. It observes a wide field of the sky to detect temporary dips in the brightness of stars, which may indicate the presence of exoplanets passing in front of them. It uses four wide-field cameras.

11.3.4 Other types of space robots

There are several other specialized robots and robotic concepts designed for specific space missions and applications. Here are a few more types of space robots:

Humanoids are robots built to look like a human. They are designed with a tall stature and equipped with advanced sensors and cameras. NASA's humanoid robots such as Robonaut 1 (R1), Robonaut 2 (R2) [28], and Valkyrie (R5) [29] are good examples. R1 and R2 are used in the International Space Station (ISS) to work along with the astronauts either autonomously or by human operators.

Swarm robotics are multiple small robots that work together autonomously to achieve a common goal. In space exploration, swarm robotics can be used for tasks such as planetary exploration, environmental monitoring, and assembling large structures in space. This is an idea proposed by Ethan Schaler [30] and is a research focus area of NASA (RFA-089).

Space tugs are robotic spacecraft designed to move satellites or other payloads from one orbit to another. Space tugs are used to optimize satellite deployment, facilitate satellite servicing, and clear space debris from crowded orbits. Automated transfer vehicle (ATV) by The European Space Agency (ESA) [31], Orion spacecraft part of NASA's Artemis program [32], and the mission extension vehicle (MEV) by Northrop Grumman [33] are some examples of space tugs.

11.4 PLANETARY EXPLORATION

11.4.1 The missions of Mars rovers

Mars exploration commenced with the launch of Sojourner [10] which is NASA's first rover to Mars. Sojourner made a successful landing on Mars on July 4, 1997, as part of the Pathfinder mission. Sojourner was equipped with cameras and scientific equipment, and was essentially a small-sized solar-powered rover. The accomplishment of this mission established a foundation for future space ventures. It also showcased the potential for autonomous exploration of Mars.

Launched in 2003 NASA's Spirit [11] and Opportunity [12] rovers landed on Mars in 2004 following in Sojourner's footsteps. These twin rovers were designed for a 90-day mission but after years of exploration, they exceeded expectations. Unfortunately, Spirit succumbed to the harsh Martian environment in 2010. Opportunity roamed Mars until 2018 making valuable discoveries about the ancient history of water-rich Mars.

Launched on November 26, 2011 NASA's Curiosity [13] rover landed in the Gale crater on the Red Planet's surface in August 2012. The arrival of Curiosity was a true technological triumph. It is equipped with the most advanced scientific instruments ever sent to another planet (17 cameras and a robotic arm), allowing scientists to explore the Martian rocks, soil, and atmosphere in unprecedented detail. Curiosity still roams Mars.

The most recent additions to the Martian rover family are Perseverance [14] and Zhurong [34], landing on Mars in 2021, heralding a new era of exploration. Perseverance is not just a rover; it is a sophisticated scientific laboratory on wheels. Its mission objectives are ambitious: to search for signs of ancient microbial life, collect and store Martian samples for future return to Earth, and test technologies for future human missions to Mars. Zhurong is the first Chinese Mars rover that explored Utopia Planitia. It is powered by the sun, and it has studied geology, atmosphere, and potential past water. It travelled over 2,700 meters and found signs of recent water activity. Table 11.2 presents information about various rovers traversed on Martian surface.

11.4.2 Venus exploration

Exploration of Venus began in the early 1960s, with the Soviet Venera program [35] taking the lead. The Venera spacecraft was the first to successfully reach and land on the surface of Venus. The extreme temperatures and pressure on Venus posed immense challenges to spacecrafts. The Soviet Union's Venera explorations from the 1960s to the 1980s provided valuable data about Venus' atmosphere and surface. Venera landers took images and conducted chemical analyses. They revealed a rocky, desolate landscape full of rocks and boulders. However, these missions faced many obstacles due to Venus' hostile environment. The extreme heat and pressure on Venus' surface caused many landers to collapse within minutes, limiting full exploration capabilities.

In the 1980s, the Soviet Union's Vega mission [36] took a different approach. They deployed balloon probes, which hovered in the upper atmosphere of Venus. These probes have provided valuable data and insight into the planet's complex cloud layers and atmospheric dynamics.

Table 11.2 Mars rovers and their mission

Feature	*Sojourner (1997) [10]*	*Spirit & Opportunity (2004) [11], [12]*	*Curiosity (2012) [13]*	*Perseverance (2021) [14]*	*Zhurong (2021) [34]*
Mission	Mars Pathfinder	Mars Exploration Rovers	Mars Science Laboratory	Mars 2020	Tianwen-1
Landing site	Ares Vallis	Gusev Crater & Meridiani Planum	Gale Crater	Jezero Crater	Utopia Planitia
Dimensions (m)	0.65 × 0.9 × 2.3	1.5 × 2.3 × 1.6	2.1 × 2.8 × 2.9	2.2 × 2.9 × 3.0	1.85 × 2.4 × 3.3
Weight (kg)	10.6	185	899	1025	240
Power source	Solar panels	Solar panels	Radioisotope thermoelectric generator (RTG) + solar panels	RTG + solar panels	Solar panels
Battery	Non-rechargeable	Non-rechargeable	Rechargeable lithium-ion	Rechargeable lithium-ion	Rechargeable lithium-ion
Instruments	APXS, IMP, LRR, MET, SUMMIT	APXS, MB, Mini-TES, Pancam, Navcam, MI, RAT, Hazcams	APXS, ChemCam, SAM, MastCam, MAHLI, MARDI, RAD, REMS, DAN	SuperCam, PIXL, SHERLOC, WATSON, Mastcam-Z, ChemCam, MOXIE, MEDA, RIMFAX	MSCD, MCS, GPR, MSC, NTC
Max speed (m/s)	0.4	0.05	0.09	0.08	0.2
Obstacle clearance (cm)	13	25	50	65	20
Mission duration (Martian sols)	83	687 (Spirit), 5111 (Opportunity)	3529 (as of Oct 2023)	671 (as of Oct 2023)	1320 (as of Oct 2023)
Computing System	8-bit Intel 80C85, 512 kB RAM, 256 kB EEPROM	20 MHz PowerPC 750, 256 MB RAM, 256 MB flash memory	2 Radiation-hardened RAD750, 2 GB RAM, 256 MB flash memory	2 Radiation-hardened RAD750, 2 GB RAM, 256 MB flash memory	2 MIPS 24K V5 cores, 256 MB RAM, 512 MB flash memory

Autonomy	Limited, mostly teleoperated	Advanced, capable of autonomous navigation and obstacle avoidance	Advanced, capable of autonomous navigation, obstacle avoidance, and target selection	Advanced, capable of autonomous navigation, obstacle avoidance, target selection, and sample caching	Advanced, capable of autonomous navigation, obstacle avoidance, target selection, and drilling
Mobility	6-wheeled rocker-bogie, 0.4 m/s max speed	6-wheeled rocker-bogie, 0.05 m/s max speed	6-wheeled rocker-bogie, 0.09 m/s max speed	6-wheeled rocker-bogie, 0.08 m/s max speed	6-wheeled rocker-bogie, 0.2 m/s max speed
Science focus	Basic geology and meteorology	Geology, mineralogy, and potential for past life	Geology, mineralogy, chemistry, and potential for past or present life	Geology, mineralogy, chemistry, and potential for past or present life, sample collection	Geology, mineralogy, potential for past water, and ice detection
Major discoveries	Evidence of past water, diverse geology	Past water activity, diverse geology, potential for microbial life	Ancient water, potential for microbial life	Evidence of past life, potential for ongoing microbial life	Evidence of past water, ice beneath the surface
Major technical achievements	First successful robotic rover on Mars	First long-duration Mars rovers, extensive geological and atmospheric data	First rover to use laser ablation and a drill, advanced autonomous capabilities	First rover to collect and store Martian rock cores, first on-site oxygen production experiment	First Chinese rover on Mars, ground-penetrating radar for subsurface exploration

In the early 1990s, NASA's Magellan [37] spacecraft mapped Venus surface in detail. This revolutionized human understanding of Venus. Magellan used radar to penetrate dense clouds and to create high-resolution maps of Venus' terrain. The mission revealed vast volcanic plains, vast rift valleys, and thousands of volcanoes, suggesting an active geological history. Magellan's data also suggests the existence of ancient river valleys and possible tectonic activity, raising interesting questions about Venus' past.

Launched on November 9, 2005, Venus Express [38] was a European Space Agency (ESA) spacecraft mission to study Venus. The spacecraft conducted observations and experiments from orbit around Venus for more than eight years. Venus Express investigated Venus' atmosphere, ground surface, clouds, and the interactions between the solar wind and Venus' atmosphere.

Although the 2010 attempt failed, the Japanese space probe Akatsuki [39] successfully entered orbit around in Venus in 2015. Akatsuki carries a suite of instruments to study Venus' atmosphere and weather patterns.

NASA's planned Venus Emissivity, Radio Science, InSAR, Topography, and Spectroscopy (VERITAS) mission [40] will map the surface of the Venus in high resolution. This is expected to provide insights into Venus' geological evolution. In addition, NASA's Deep Atmosphere Venus Investigation of Noble gases, Chemistry, and Imaging Plus (DAVINCI+) [41] is aimed at understanding Venus' atmosphere and possible signs of past or present life.

11.4.3 Moon exploration

The journey to explore the Moon began with the space race in the 1960s. On July 20, 1969, NASA's Apollo 11 mission [42] accomplished the impossible: humans set foot on the moon. Neil Armstrong and Buzz Aldrin became the first humans to set foot on moon. After Apollo 11, NASA conducted a series of Apollo missions until 1972 in which a total of 12 astronauts explored the lunar surface. These missions provided valuable scientific data such as lunar samples, seismic measurements, and detailed photographs. Lunar samples brought back by the Apollo missions provided important insights into the Moon's composition, revealing the presence of water molecules and traces of volatile elements. These results challenged previous assumptions and raised new questions about the history of the Moon and the processes forming its surface.

In 2009 NASA's LCROSS mission [43] discovered water ice in a permanently shadowed crater near the Moon's south pole. This discovery had far-reaching consequences, indicating that the Moon could be used for future exploration and human settlement. The findings are also significant.

Over the past few decades, the European Space Agency's SMART, India's Chandrayaan, China's Chang'e series, and NASA's Lunar Reconnaissance Orbiter (LRO) have mapped the Moon's surface in unprecedented detail.

These missions have identified regions with high concentrations of water ice, investigated the Moon's complex geological features, and improved human understanding of the Moon's evolution. Launched in 2019 China's Chang'e-4 mission reached a historic milestone with an unprecedented landing on the Moon's far side. Chang'e-4 mission's Yutu-2 rover continues to explore this uncharted territory, conducting scientific experiments and studying the Moon's geology in detail.

New frontiers in lunar research are being forged by commercial ventures and international collaborations. NASA's Artemis mission aims to return humans to the lunar surface. Artemis seeks to establish a sustainable human presence on the Moon and use it as a springboard for future human missions to Mars and beyond. Commercial companies such as SpaceX [44] and Blue Origin [45] are working to develop lunar landers and habitats, fostering a new era of space entrepreneurship.

11.5 ASTEROID AND COMET EXPLORATION

The exploration of asteroids began with flyby missions, such as NASA's Galileo spacecraft [34], which passed by two asteroids, Gaspra and Ida, in the 1990s. These missions provided the first close-up images of asteroids and revealed their irregular shapes and diverse surface features. Galileo continued its travel to Jupiter.

NASA's OSIRIS-REx [35] (Origins, Spectral Interpretation, Resource Identification, Security, Regolith Explorer) mission, launched in 2016, represents a groundbreaking endeavor in asteroid exploration. OSIRIS-REx successfully rendezvoused with the near-Earth asteroid Bennu in 2018. The mission objectives are sample collection, understanding the early solar system, planetary defense, and resource utilization. Over the course of its mission, OSIRIS-REx conducted detailed surveys, mapping Bennu's surface and studying its composition. In a historic feat, the spacecraft executed a touch-and-go (TAG) maneuver in October 2020, collecting samples from Bennu's surface. These samples were returned to Earth in 2023. The spacecraft did not land, it continued its journey to explore another asteroid – Apophis (OSIRIS-APEX).

The spacecraft employed an array of instruments as listed below to study Bennu:

- OCAMS (OSIRIS-REx Camera Suite) captured high-resolution images of Bennu's surface.
- OTES (OSIRIS-REx Thermal Emission Spectrometer): measured the temperature of Bennu's surface, providing information about the composition of surface materials.
- OVIRS (OSIRIS-REx Visible and Infrared Spectrometer): analyzed the spectral properties of Bennu's surface, aiding in the identification of minerals and organics.

- OLA (OSIRIS-REx Laser Altimeter): mapped Bennu's topography by bouncing laser beams off its surface, helping to understand the asteroid's shape and structure.
- TAGSAM (Touch-And-Go Sample Acquisition Mechanism): the specialized robotic arm designed to perform the touch-and-go maneuver for sample collection.
- TAG (Touch-And-Go) Maneuver.

The most important phase of the OSIRIS-REx mission was the TAG maneuver. TAG maneuver was conducted on October 20, 2020. That day the spacecraft descended to Bennu's surface and made a brief contact for about six seconds to collect regolith samples. A burst of nitrogen gas was released to stir up the regolith, which became trapped in the TAGSAM head. The spacecraft's engines were ignited shortly after the sample collection to exit Bennu. OSIRIS-REx delivered the samples to Earth on September 24, 2023. The spacecraft continued its journey to explore another asteroid- Apophis (OSIRIS-APEX).

Launched in 2003, the first Hayabusa mission [36] successfully returned samples from the asteroid Itokawa in 2010. Based on this success, the Hayabusa2 mission [37] was launched in December 2014. This mission aimed to explore the near-Earth asteroid Ryugu and return samples to Earth. The mission's goal was to collect samples, understand the asteroid's composition, analyze its subsurface, and discover the ways and means to interact or potentially alter the course of an asteroid. Hayabusa2 collected samples from the surface of Ryugu in 2019 and returned to Earth in December 2020. The spacecraft used many of the instruments listed below to study Ryugu:

- Small Carry-on Impactor (SCI), a 2 kg copper nugget for creating artificial crater on Ryugu to study its subsurface material.
- Near infrared spectrometer (NIRS) and thermal infrared imager (TIR) for remote sensing the asteroid.
- Small rovers to descend on the surface to study it.
- Small lander (MASCOT) containing four observation devices (MicrOmega, MAG, CAM, and MARG) to land on and study the asteroid.

11.6 SATELLITE SERVICING AND REPAIR

Satellites in orbit face a finite operational lifespan determined by variables such as fuel consumption, orbital positioning, and external forces. Prior to satellite deployment, careful consideration is given to establishing the anticipated operational duration. A pivotal factor influencing a satellite's endurance is its capacity for periodic repairs in space. This critical function is facilitated by robots, with leading entities such as Commercial Made in

Space, Inc., Northrop Grumman Innovation Systems, and NASA providing specialized satellite repair services, as highlighted in [46].

The trajectory of satellite repairs in space is contingent upon diverse robotic technologies. Ongoing initiatives, including the evolution of intelligent robots tailored for satellite servicing, exemplify sustained progress in this domain. Jiang et al. [47] delve extensively into the utilization of robots to support satellites in space. The implementation of cutting-edge robotic technologies ensures a continual enhancement in the quality of service delivered to satellites over time.

A widely favored approach for conducting satellite repair and servicing involves the deployment of robotic arms and manipulators. These space robots operate in orbit, where satellites carry out their functions, and are capable of performing various tasks for astronauts, including repairing, and maintaining satellites in their orbital positions. The advanced technology of self-operating robotic arms enables them to execute repair tasks, adjust positions, and conduct inspections. While existing robotic arms may or may not meet the requirements of specific space missions, the need for tailored robotic arms arises, leading to the design and development of specialized robotic arms for use in space missions. Numerous companies have undertaken the development of such robotic arms for space missions. In [48], Won et al. discuss emerging trends in the development of space robotic arms and manipulators specifically designed for space missions.

The pioneering advancement in the design and development of robotic arms began with the Canadarm series, initiated by the Canadian Space Agency in 1981. This series remained operational for over 30 years, successfully executing numerous missions. Currently, the Canadarm 3 is in the development phase, and there are plans to provide it to NASA in 2024 for a Lunar Gateway space station. In 2008, the Canadian Space Agency, in collaboration with MDA, initiated the Dexter project, featuring a robotic system with double arms used by astronauts during spacewalks. Another notable robotic arms project launched in 2021 is the European Robotic Arm, developed by the Dutch space agency. These robotic arms were employed for moving small payloads between the interior and exterior of space stations. Another project, the Kraken robotic arm, designed for manufacturing and assembly space station missions, proved particularly useful for small satellites, although testing for this project is still ongoing.

In the realm of space exploration, one of the prominent missions is Mars exploration. The initial design and deployment of robotic arms for this Mars mission took place in 2004 with the Mars Exploration Rover robotic arms, and more recently, in 2020, the Mars Science Laboratory robotic arm was deployed. The robotic arm, on the Mars Exploration Rover is utilized to interact with rocks and soil on Mars. Similarly the Mars Science Laboratory's robotic arm has played a role in missions that explore paths and collect samples of soil and rocks. Although there have been advancements and research in this field further exploration is necessary to gain an understanding of how robotics can assist in space exploration.

Several ongoing projects involve the utilization of arms for repair and servicing purposes. Examples include xLink, Space Infrastructure Dexterous Robot (SPIDER) and Lunar Lightweight Surface Manipulator System. Moreover, robotic arms like FREND and the K Space Robot Manipulator are currently being employed for space repairs and servicing. Numerous researchers are actively working on finding solutions to address challenges encountered during repairing and servicing space missions.

The application of space robots provides a solution to the challenges faced when repairing space missions. Notably using space robots for satellite repairs can help mitigate space debris issues while extending the lifespan of satellites. While there are advantages associated with using space robots, for satellite repairs numerous challenges also arise. Mobility is severely restricted in space and communication problems add complexity to repair tasks. Precisely positioning tool equipment is another substantial challenge in satellite repairs using robots. Costs, required skills, and expertise further compound these challenges. Despite these obstacles, the benefits of satellite repairs through robots outweigh the challenges, presenting multiple opportunities for the future.

11.7 SPACE MINING AND RESOURCE UTILIZATION

Many companies and organizations are conducting research and mining operations on Earth to discover valuable hidden resources. Nowadays, mining and resource exploration research is not limited to Earth alone; it has already expanded into space. There are numerous celestial bodies in space that may contain valuable resources for utilization. Robots play a vital role in extending the research into mining these resources. The discovery of these valuable resources will open various paths for human life on the Moon and Mars. It is essential to understand how robotics can contribute to space mining and resource utilization research.

In a study by Just et al. [49], they discuss a review of excavation techniques for Lunar In Situ Resource Utilization (ISRU) and make recommendations for future excavation experiments. Given the possibility of existing water on the Moon, robots equipped with drilling tools can be used to search for existing ice. Robots can also provide heating facilities to transform identified solid resources into a liquid state. Additionally, robots are essential for storage and transportation within space satellites. They assist in sample collection and site exploration using cameras and specialized sensors. Robots equipped with environmental sensors can help locate valuable minerals in lunar space.

In [50], Meurisse and Carpenter discussed the past, present, and future rationale for space resource utilization. They researched how the exploration of space resources began, what has been successful, and the challenges encountered. Valuable resources hidden on the Moon, Mars, and asteroids are the focus of various space missions that utilize emerging robotic technology.

One of the most critical resources to support life is water, and many space missions are employing robots to search for this resource. Some space missions are also working to develop on-site propellants, which can provide long-term stability for satellites in space and reduce transportation costs for these resources. Ongoing research is still focused on enhancing the utilization of current robotic tools to discover valuable resources on different celestial bodies in the universe.

11.8 HUMAN–ROBOT COLLABORATION

Robots can perform various tasks on their own, but their interaction with humans makes them more powerful tools for research and exploration. In space, where robots play a crucial role in research and other operations, human interaction ensures their success in completing their allocated missions. Remotely operable robots, controlled by humans, can be used for space exploration. Space robots are used for data collection, data analysis, and decision-making based on their learned environment. Humans can navigate robots toward their missions from Earth, making human collaboration essential for space missions.

Human control of robots in space is essential for tasks like astronaut assistance, telepresence, and planetary exploration. These are key uses of human-interacted robots in space stations. However, there are challenges in this regard, including gravity issues, communication problems, radiation exposure, and high temperatures in space. Some agencies are still conducting research to develop advanced, high-tech human-interacted robots that can provide additional benefits for space missions.

11.9 ETHICAL AND ENVIRONMENTAL CONSIDERATIONS

Nowadays, scientists and researchers are exploring the mineral and valuable resources in space by conducting various experiments. It is really important to understand the consequences of these space experiments. The most significant issue with these experiments is the ethical and environmental considerations. In this section, we will discuss the major ethical and environmental issues in space exploration and also discuss how robotics can provide countermeasures for these issues.

It is first required to understand how ethical and environmental issues are related to space exploration experiments. Space debris is the biggest concern in space exploration. This debris is left in space during space repairs or other experiments. According to ethical concerns, the space repair station operation should have ethical responsibility to minimize this debris. Another significant ethical concern is about the security of data. During the space mission, satellites keep sending data to ground control stations that help scientists and researchers understand space exploration experiments. The privacy of this transmitted data should be maintained during transmission. The next ethical issue is related to the increasing number of space stations or settling in space, which can cause a high risk of collision of those satellites. There should be ethical guidelines for traffic management in orbital space that should be in collaboration with the International Space Organization to overcome these issues. It also needs to be considered regarding the end of life of any satellite in space. There should be proper guidelines on how these satellites can be moved to graveyard orbits.

Space exploration also leads to some major environmental issues, such as a satellite consuming massive energy during the launch issue, impacting a lot on the environment. Orbital congestion due to the increasing number of satellites year by year also impacts the environment. If these are not properly managed, it will cause satellite collisions. Satellite production is a huge project, and it not only costs a considerable amount but also uses a lot of materials, which is directly related to environmental issues. Despite having the above issues in space exploration, we cannot stop our research in these areas.

In reference [51], Miraux et al. discuss the environmental sustainability of future space activities. They carried out a life cycle assessment from 2022 to 2050, examining the environmental impact within the space sector. Lawrence et al. also address various issues related to space environmentalism in their research [52]. Therefore, it is important to balance these space exploration research experiments by considering all these issues.

Understanding how space robots play a vital role in mitigating ethical and environmental issues is fascinating. One approach to overcoming these issues in the future is through the implementation of an ethical code for space environmental development, as discussed by Williamson in [53]. Space robots can also play an important role in mitigating these issues. Robotic arms can be used to remove space debris during space missions. Space robots equipped with highly advanced sensors and cameras can detect other satellites in the same space region, helping to avoid collisions, especially in cases of high space traffic.

Advanced artificial intelligence and robotic technology can be very helpful in autonomous satellite navigation, reducing the risk of collisions in the space environment. With the help of autonomous self-navigating satellites using artificial intelligence and robotic technology, a safe distance can be maintained between satellites, promoting a collaborative approach for the international space organization.

Robotic systems also ensure secure communication between satellites and ground stations, maintaining privacy for data transmission. As discussed earlier in Section 11.6, space robots are also useful for repairing satellites, extending their lifespan in the space station. Robotics and reusable spacecraft designs can be used to reduce energy consumption. Repair operations can also help reduce the number of space launches, mitigating environmental issues.

Space robots are also useful for upgrading new technology without complete replacement and can enhance the efficiency of solar panel optimization. Overall, their use in space helps overcome ethical and environmental issues. We can say that smart space robots, with their advanced tools and intelligent capabilities, assist us in performing space activities in a responsible and safe manner. As we improve our use of these robots, they will likely become increasingly important for solving problems and maintaining the cleanliness of our space. As technology advances, using robots to assist satellites will likely play a significant role in making space activities environmentally friendly and compliant with regulations.

11.10 FUTURE PROSPECTS AND EMERGING TECHNOLOGIES

Due to advanced technology and development, the future of space robots looks very promising. Researchers are exploring numerous areas that will undoubtedly be available in the future. Some researchers are focusing on autonomous operations for space satellites. With the assistance of space robots and advanced artificial intelligence technology, satellites can execute their missions without human intervention. They can make decisions independently in the space environment for space exploration.

Another future project achievable with the help of robotics is the deployment of a swarm of robots in space. Space robots can coordinate in the space environment and perform various useful operations through collaboration. Another emerging technology we can expect to see in the future space environment is 3D printing. With the aid of this emerging technology, a 3D robotic printer can construct tools and parts used in space missions, eliminating the need to transport those parts from Earth.

In the future, trained robots can work alongside astronauts to execute various space missions and conduct new experiments in space exploration. Advanced robotics will also provide the opportunity to enhance mobility in the space environment. For example, with enhanced navigation and traverse-capable robotic arms, advanced rovers can perform complex space missions.

In space missions, human intervention often poses challenges during communication. We can envision a future where autonomous robots working on space missions will increase long-term capabilities with improved communication. Another area where scientists and researchers are working on for space exploration, and which will likely become possible in the future, is the use of artificial intelligence and machine learning. These technologies will enhance navigation and provide a better way for data analysis, improving overall efficiency in space missions.

Robotic and emerging technologies will provide advanced functionality for future space missions. Rovers and landers will be able to perform more complex missions in the future with their advanced technology and features. The research in space missions knows no limits, and we anticipate better opportunities with the help of advanced robotics in the future.

11.11 CONCLUSION

The application of robotics is not limited to operations performed on Earth; it extends far beyond and is being used in space exploration. In this chapter, we provide the history and classification of space robots and their current applications in space exploration. As most space missions, whether to the moon or Mars, aim to discover valuable resources such as water on their surfaces to facilitate human life, countries conduct research and deploy satellites in space for these missions. The use of robotics enhances mission success by deploying robots capable of exploring space, robotic arms for repairing, and collecting samples from different planetary missions.

This chapter comprehensively covers all aspects of space robots, their applications, and future technologies that can contribute to their missions. While these advanced technologies show promise in resolving complex issues, further research is needed to advance and enhance their technological capabilities for space missions. The advanced technology of space robotics, coupled with the use of AI, enables autonomous space missions without human intervention. These AI autonomous space missions elevate the possibilities in space exploration.

The chapter encompasses all information about space robotics and provides a future research direction that can prove more beneficial in space exploration missions with the integration of AI and robotics.

REFERENCES

[1] JPL, "JPL," NASA Jet Propulsion Laboratory (JPL). Accessed: Dec. 04, 2023. [Online]. Available: www.jpl.nasa.gov/

[2] A. Zak, "Russia's unmanned missions to Venus." Accessed: Dec. 04, 2023. [Online]. Available: www.russianspaceweb.com/spacecraft_planetary_venus.html

[3] P. J. Westwick, *Into the Black: JPL and the American Space Program, 1976-2004*. Yale University Press, 2008. [Online]. Available: https://books.google.com.au/books?id=YXSxmgVLhskC

[4] JPL-Surveyor1, "Surveyor 1 – Moon Missions – NASA's Jet Propulsion Laboratory," NASA Jet Propulsion Laboratory (JPL). Accessed: Dec. 04, 2023. [Online]. Available: www.jpl.nasa.gov/missions/surveyor-1

[5] W. R. Corliss, *The Viking Mission to Mars*. Scientific and Technical Information Office, National Aeronautics and Space Administration, 1976.
[6] JPL-Voyager1, "Voyager 1 – Interplanetary Missions – NASA Jet Propulsion Laboratory," NASA Jet Propulsion Laboratory (JPL). Accessed: Dec. 04, 2023. [Online]. Available: www.jpl.nasa.gov/missions/voyager-1
[7] JPL-Voyager2, "Voyager 2 – Interplanetary Missions – NASA Jet Propulsion Laboratory," NASA Jet Propulsion Laboratory (JPL). Accessed: Dec. 04, 2023. [Online]. Available: www.jpl.nasa.gov/missions/voyager-2
[8] NASA-Stardust, "Stardust – NASA's Comet Sample Return Mission." Accessed: Dec. 04, 2023. [Online]. Available: https://solarsystem.nasa.gov/stardust/home/index.html
[9] ESA-Rosetta, "Rosetta." Accessed: Dec. 04, 2023. [Online]. Available: www.esa.int/Science_Exploration/Space_Science/Rosetta
[10] JPL-Sojourner, "Mars Pathfinder – Mars Missions – NASA Jet Propulsion Laboratory," NASA Jet Propulsion Laboratory (JPL). Accessed: Dec. 04, 2023. [Online]. Available: www.jpl.nasa.gov/missions/mars-pathfinder-sojourner-rover
[11] JPL-Spirit, "Mars Rover Spirit – Mars Missions – NASA Jet Propulsion Laboratory," NASA Jet Propulsion Laboratory (JPL). Accessed: Dec. 04, 2023. [Online]. Available: www.jpl.nasa.gov/missions/mars-exploration-rover-spirit-mer-spirit
[12] JPL-Opportunity, "Mars Rover Opportunity – Mars Missions – NASA Jet Propulsion Laboratory," NASA Jet Propulsion Laboratory (JPL). Accessed: Dec. 04, 2023. [Online]. Available: www.jpl.nasa.gov/missions/mars-exploration-rover-opportunity-mer
[13] JPL-Curiosity, "Mars Science Laboratory Curiosity Rover – Mars Missions – NASA Jet Propulsion Laboratory," NASA Jet Propulsion Laboratory (JPL). Accessed: Dec. 04, 2023. [Online]. Available: www.jpl.nasa.gov/missions/mars-science-laboratory-curiosity-rover-msl
[14] JPL-Perseverance, "Mars 2020 – NASA Jet Propulsion Laboratory," NASA Jet Propulsion Laboratory (JPL). Accessed: Dec. 04, 2023. [Online]. Available: www.jpl.nasa.gov/missions/mars-2020-perseverance-rover
[15] S. Kassel, "Lunokhod-1 Soviet Lunar Surface Vehicle," *Rand*, p. 31, 1971.
[16] C. Li *et al.*, "The Chang'e 3 mission overview," *Space Sci. Rev.*, vol. 190, no. 1–4, pp. 85–101, Jul. 2015, doi: 10.1007/s11214-014-0134-7
[17] L. Ding *et al.*, "A 2-year locomotive exploration and scientific investigation of the lunar farside by the Yutu-2 rover," *Sci. Robot.*, vol. 7, no. 62, p. eabj6660, Jan. 2022, doi: 10.1126/scirobotics.abj6660
[18] S. Majumdar, "ISRO's Chandrayaan–3 mission a success," *Vayu Aerosp. Def. Rev.*, vol. 5, no. 5, pp. 29–31, Oct. 2023.
[19] Anonymous, "Chandrayaan 2," *Vayu Aerosp. Def. Rev.*, vol. 5, no. 5, p. 4, Oct. 2019.
[20] NASA-Viper, "VIPER – NASA Science." Accessed: Dec. 06, 2023. [Online]. Available: https://science.nasa.gov/mission/viper/
[21] S. Mallapaty, "China has landed its first rover on Mars – here's what happens next," *Nature*, vol. 593, no. 7859, pp. 323–325, May 2021, doi: 10.1038/d41586-021-01301-7

[22] Hubble, "Hubble Space Telescope – NASA Science," Hubble Space Telescope. Accessed: Dec. 06, 2023. [Online]. Available: https://science.nasa.gov/mission/hubble/
[23] NASA, "NASA's First Stellar Observatory, OAO 2, Turns 50 – NASA." Accessed: Dec. 06, 2023. [Online]. Available: www.nasa.gov/solar-system/nasas-first-stellar-observatory-oao-2-turns-50/
[24] Chandra, "Chandra – NASA Science." Accessed: Dec. 06, 2023. [Online]. Available: https://science.nasa.gov/mission/chandra/
[25] Harvard, "Chandra:: About Chandra:: Science Instruments," Chandra X-Ray Discovery. Accessed: Dec. 06, 2023. [Online]. Available: https://chandra.harvard.edu/about/science_instruments.html
[26] NASA-Fermi, "The Fermi Gamma-ray Space Telescope," NASA. Accessed: Dec. 06, 2023. [Online]. Available: https://fermi.gsfc.nasa.gov/
[27] K. Walbolt, "Transiting Exoplanets Survey Satellite (TESS)," Exoplanet Exploration: Planets Beyond our Solar System. Accessed: Dec. 06, 2023. [Online]. Available: https://exoplanets.nasa.gov/tess
[28] NASA-Robonaut2, "Robonaut2 – NASA." Accessed: Dec. 08, 2023. [Online]. Available: www.nasa.gov/robonaut2/
[29] NASA-R5, "R5 – NASA." Accessed: Dec. 08, 2023. [Online]. Available: www.nasa.gov/technology/r5/
[30] NASA-Swarm, "Swarm of Tiny Swimming Robots Could Look for Life on Distant Worlds – NASA." Accessed: Dec. 08, 2023. [Online]. Available: www.nasa.gov/directorates/stmd/niac/swarm-of-tiny-swimming-robots-could-look-for-life-on-distant-worlds/
[31] ESA-ATV, "ATV." Accessed: Dec. 08, 2023. [Online]. Available: www.esa.int/Science_Exploration/Human_and_Robotic_Exploration/ATV/
[32] NASA-Orion, "Meet NASA's Orion Spacecraft – NASA." Accessed: Dec. 08, 2023. [Online]. Available: www.nasa.gov/missions/meet-nasas-orion-spacecraft/
[33] Northrop Grumman, "Digital Engineering Triumph MEV Revitalizes Satellites in Space," Northrop Grumman. Accessed: Dec. 08, 2023. [Online]. Available: www.northropgrumman.com/space/digital-engineering-triumph-mev-revitalizes-satellites-in-space
[34] L. Ding *et al.*, "Surface characteristics of the Zhurong Mars rover traverse at Utopia Planitia," *Nat. Geosci.*, vol. 15, no. 3, pp. 171–176, Mar. 2022, doi: 10.1038/s41561-022-00905-6
[35] D. Adler, "Behind the Iron Curtain: The Soviet Venera program," Astronomy Magazine. Accessed: Dec. 21, 2023. [Online]. Available: www.astronomy.com/science/the-venera-program-interplanetary-probes-from-behind-the-iron-curtain/
[36] A. Kuzmin, "VEGA Mission," Planetary Data Image Processing Lab. Accessed: Dec. 22, 2023. [Online]. Available: http://arc.iki.rssi.ru/IPL/vega.html
[37] NASA-Magellan, "Magellan – NASA Science," NASA. Accessed: Dec. 22, 2023. [Online]. Available: https://science.nasa.gov/mission/magellan/
[38] ESA-Venus Express, "Venus Express," The European Space Agency. Accessed: Dec. 22, 2023. [Online]. Available: www.esa.int/Science_Exploration/Space_Science/Venus_Express

[39] ISAS-Akatsuki, "Venus Climate Orbiter AKATSUKI web site," Venus Climate Orbiter AKATSUKI. Accessed: Dec. 22, 2023. [Online]. Available: http://akatsuki.isas.jaxa.jp/en/
[40] NASA-Veritas, "VERITAS," NASA Jet Propulsion Laboratory (JPL). Accessed: Dec. 22, 2023. [Online]. Available: www.jpl.nasa.gov/missions/veritas
[41] NASA-Future, "NASA Selects Four Possible Missions to Study the Secrets of the Solar System," NASA Jet Propulsion Laboratory (JPL). Accessed: Dec. 22, 2023. [Online]. Available: www.jpl.nasa.gov/news/nasa-selects-four-possible-missions-to-study-the-secrets-of-the-solar-system
[42] NASA-Apollo11, "The Apollo Experiment That Keeps on Giving," NASA Jet Propulsion Laboratory (JPL). Accessed: Dec. 22, 2023. [Online]. Available: www.jpl.nasa.gov/news/the-apollo-experiment-that-keeps-on-giving
[43] NASA-Lcross, "Lunar Impact May Impact Lunar Science For Years To Come," NASA Jet Propulsion Laboratory (JPL). Accessed: Dec. 22, 2023. [Online]. Available: www.jpl.nasa.gov/news/lunar-impact-may-impact-lunar-science-for-years-to-come
[44] SpaceX, "SpaceX," SpaceX. Accessed: Dec. 22, 2023. [Online]. Available: www.spacex.com
[45] Blue Origin, "Home," Blue Origin. Accessed: Dec. 22, 2023. [Online]. Available: www.blueorigin.com
[46] J. P. Davis, J. P. Mayberry, and J. P. Penn, "On-orbit servicing: Inspection repair refuel upgrade and assembly of satellites in space," *Aerosp. Corp. Rep.*, p. 25, 2019. Available: https://csps.aerospace.org/sites/default/files/2021-08/Davis-Mayberry-Penn_OOS_04242019.pdf
[47] Z. Jiang, X. Cao, X. Huang, H. Li, and M. Ceccarelli, "Progress and development trend of space intelligent robot technology," *Space Sci. Technol.*, vol. 2022, p. 2022/9832053, Jan. 2022, doi: 10.34133/2022/9832053
[48] D. Won, B.-R. So, and H.-D. Kim, "A Survey of Space Robotic Manipulator," *J. Space Technol. Appl.*, vol. 2, no. 4, pp. 257–267, Nov. 2022, doi: 10.52912/jsta.2022.2.4.257
[49] G. H. Just, K. Smith, K. H. Joy, and M. J. Roy, "Parametric review of existing regolith excavation techniques for lunar In Situ Resource Utilisation (ISRU) and recommendations for future excavation experiments," *Planet. Space Sci.*, vol. 180, p. 104746, Jan. 2020, doi: 10.1016/j.pss.2019.104746
[50] A. Meurisse and J. Carpenter, "Past, present and future rationale for space resource utilisation," *Planet. Space Sci.*, vol. 182, p. 104853, Mar. 2020, doi: 10.1016/j.pss.2020.104853
[51] L. Miraux, A. R. Wilson, and G. J. Dominguez Calabuig, "Environmental sustainability of future proposed space activities," *Acta Astronaut.*, vol. 200, pp. 329–346, Nov. 2022, doi: 10.1016/j.actaastro.2022.07.034
[52] A. Lawrence *et al.*, "The case for space environmentalism," *Nat. Astron.*, vol. 6, no. 4, pp. 428–435, Apr. 2022, doi: 10.1038/s41550-022-01655-6
[53] M. Williamson, "Space ethics and protection of the space environment," *Space Policy*, vol. 19, no. 1, pp. 47–52, Feb. 2003, doi: 10.1016/S0265-9646(02)00064-4

Index

FSC
www.fsc.org
MIX
Papier | Fördert gute Waldnutzung
FSC® C083411

Kolibri 360 GmbH
Ferdinand-Jühlke-Straße 7
99095 Erfurt, Deutschland
produktsicherheit@kolibri360.de